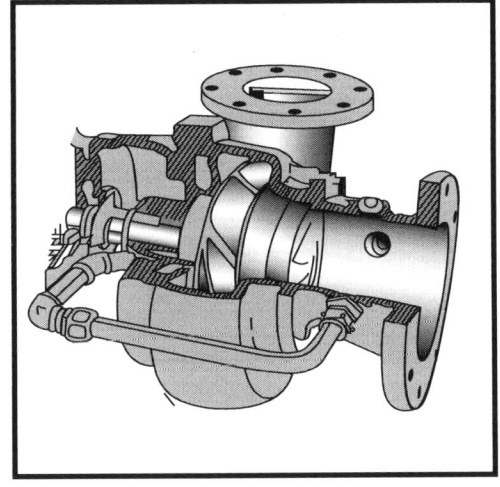

# Pumps & Pumping Systems

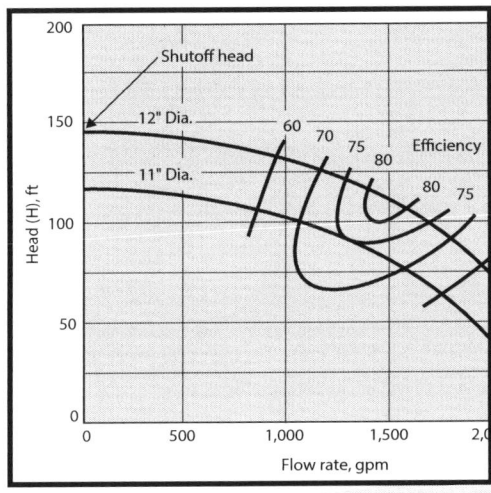

October 2015

Edited by
Robert D. von Bernuth,
PhD, PE, CID, CLWM, CIC

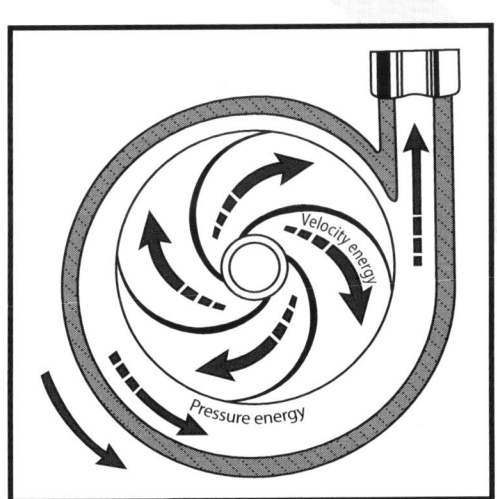

## About IA

The Irrigation Association is the leading membership organization for irrigation companies and professionals. Together with its members, IA is committed to promoting efficient irrigation technologies, products, and services and to long-term sustainability of water resources for future generations. IA serves its members and the industry by

- improving industry proficiency through continuing education.
- recognizing and promoting experience and excellence with professional certification.
- influencing water-use policy at the local, state, and national levels.
- ensuring industry standards and codes reflect irrigation best practices.
- providing forums that promote innovative solutions and efficient irrigation practices and products.

For more information, visit www.irrigation.org.

## About the Foundation

The Irrigation Foundation advances the irrigation industry using education that both attracts people into the profession and ensures ongoing professional success.

Founded in 1980, the Foundation is a nonprofit, tax-exempt 501(c)(3) corporation whose operations are funded entirely by tax-deductible contributions and program revenue.

For more information, visit www.irrigationfoundation.org.

**Irrigation Association**
8280 Willow Oaks Corporate Drive
Suite 400
Fairfax, VA 22031-4507
Tel: 703.536.7080
Fax: 703.536.7019
info@irrigation.org
www.irrigation.org

© 2015 by the Irrigation Association
All rights reserved. Published 2015

No part of this book may be reproduced in any form or by any means without permission in writing from the Irrigation Association.

This Irrigation Association publication is informational and advisory only. The Irrigation Association assumes no responsibility for results attributable to this book. This publication is designed to provide accurate and authoritative information in regard to the subject matter covered. It is presented with the understanding that the publisher is not engaged in rendering professional service.

ISBN: 978-1-935324-30-0

**October 2015** (j15)

# Table of Contents

## Chapter 1 — Introduction
Background...........................................................................................1
What Is a Pump?....................................................................................1
When Is a Pump Necessary?.................................................................2

## Chapter 2 — Hydraulics Review
Energy....................................................................................................5
Continuity..............................................................................................7

## Chapter 3 — Types of Pumps
Types of Pumps Used in Irrigation.....................................................11
Centrifugal Pumps...............................................................................11
Vertical Turbine Pumps.......................................................................15
Submersible Pumps............................................................................20
Jet Pumps.............................................................................................22
Positive Displacement Pumps............................................................22

## Chapter 4 — Basics of Pumps: Force, Work, and Power
Work.....................................................................................................27
Power...................................................................................................27

## Chapter 5 — Pump Performance
Selection Criteria.................................................................................31
Design Flow Rates...............................................................................32
Design Pressure — Total Dynamic Head...........................................35
Net Positive Suction Head..................................................................39
Submergence......................................................................................39
Total Dynamic Suction Lift..................................................................41
Cavitation.............................................................................................42
Pressure Within the Pump..................................................................43
Pump Curves.......................................................................................44
Centrifugal Pump Curves....................................................................45
Vertical Turbine Pump Curves............................................................47
Pump Selection...................................................................................49

## Chapter 6 — Families of Curves
Impeller Diameter and Speed Families of Curves .................................53
Affinity Laws ...............................................................55
Changing the Pump Speed....................................................56
Multiple Pump Stations .....................................................58
Pumps in Series.............................................................58
Pumps in Parallel...........................................................59

## Chapter 7 — Pump Selection
Matching Selection Criteria..................................................63
Total Dynamic Head.........................................................63
Selection Procedure.........................................................63
System Requirements .......................................................64
Pumpage ..................................................................67
Net Positive Suction Head ...................................................68
Reading Pump Curves.......................................................71
Computer Programs for Pump Selection ......................................76
Smart Phone Apps for Pump Selection........................................81

## Chapter 8 — Operating Point for Pumps
Pump and System Curve Together.............................................85
System Curve ..............................................................85
Changes in the System ......................................................89

## Chapter 9 — Power Units
Power Unit Options .........................................................93
Power Plant Efficiency......................................................102
Comparing Energy Sources..................................................103

## Chapter 10 — Automation and Control
Pumping System Controls...................................................107
Automation............................................................... 108
Integrating Pump Stations with Irrigation Controllers .........................111
Variable Frequency Drives..................................................111
VFD Potential Problems ....................................................113

## Chapter 11 — Operation and Maintenance

Using VFDs to Lower Energy Costs ................................................121
Using VFDs to Improve the Operation of the Pumping Station ..................122
Prepackaged Pump Stations with VFD Control ..................................122
Motor Management and Maintenance...........................................123
Overheating ..................................................................123
Preseason and Postseason Checklists ..........................................124
Routine Inspection and Maintenance ..........................................125

## Chapter 12 — Irrigation Pumping Costs

Comparing Alternatives .......................................................131
Fixed Costs ..................................................................131
Demand or Other Fixed Energy Charges .......................................132
Variable Costs................................................................132
Operating Cost ...............................................................133

## Appendix A — Capital Recovery Factors ............................ 141

## Appendix B — Answers to Practice Questions...................... 143

# Chapter 1

# Introduction

## Learning Objectives

The following objectives are the focus of chapter 1:
- define what a pump is
- determine when a pump is needed

## Background

Pumps are widely used in irrigation, and a basic understanding of pumps will significantly help anyone involved in the industry. While there are a wide variety of pumps available, centrifugal pumps are most commonly used in irrigation. This course is an introduction to centrifugal pumps. The topics covered are as follows:

- hydraulics review
- energy equation
- basics of pumps
- types of pumps
- pump performance
- families of pump curves
- operating point for pumps
- pump selection
- power units
- automation and control
- operation and maintenance
- pumping costs

## What Is a Pump?

A pump is a device that takes input energy (likely from a motor or engine) and converts it into pressure head and velocity so that it can be used to move water from one point to another and to pressurize irrigation systems. It can also be used to precisely pump small quantities of specific chemicals used in the systems.

# When Is a Pump Necessary?

## *Unpressurized Source*

If the water source isn't pressurized or lacks adequate pressure, then a pump is needed. Drawing water from unpressurized sources such as wells, ponds, or streams requires use of a pump. Normally the pump is used to pressurize the irrigation system. But, at times the water may be pumped to a higher elevation, and the pressure resulting from the elevation is used to power the system. There are situations where the pump is used simply to transfer or lift water. A windmill-powered pump for watering livestock lifts water from a well or other unpressurized source to a tank accessible to the livestock, but it isn't necessary to pressurize the delivery. However, for pressurized irrigation systems, the pump not only delivers the water to the location, but it also provides the pressure for the system to operate. When a pump is used to deliver water from a well, pond, or stream to a sprinkler or drip system, it lifts the water to the proper elevation and delivers enough pressure for the system to operate.

## Pressurized Source

Water is often available under some pressure, but in many situations the pressure isn't adequate for the system to operate properly. Potable water delivered to a residence usually has enough pressure to operate small irrigation systems, but the pressure may not be adequate for a larger system. Similarly, the pressure at the end of a center pivot may not be adequate for the operation of an end gun. In both situations, water is delivered under pressure, but the pressure is too low to ensure proper system operation. Therefore, it is necessary to install a booster pump. The booster pump will raise the pressure to the proper operating range for the system.

In summary, a pump is needed (1) when the source is unpressurized and located below where it will be used, or (2) if the pressurized source does not supply adequate pressure. The amount of pressure needed depends upon system design.

# Practice Questions

1. What is a pump?

   _____

   _____

   _____

   _____

   _____

2. What are two conditions when a pump is needed?

   (1.) _____

   (2.) _____

**Chapter 1: Introduction**

# Chapter 2

# Hydraulics Review

## Learning Objectives

The following objectives are the focus of chapter 2:
- understand the energy equation and how it applies to pumps
- understand the continuity equation and how it applies to pumps

## Energy

There are two kinds of energy. Kinetic energy is the energy of a particle in motion, usually expressed in terms of velocity. The second kind of energy is potential energy. The term potential energy is used to describe the energy of a body (e.g., volume of water) due to its position or configuration within a force field. For purposes of this workbook, the force field is gravity. So, the water can have potential energy due to being located higher (against the force of gravity) or due to being constrained or pushed on. The latter could be due to a force being applied to a movable piston in a cylinder from which the water cannot leak. An example is a cylinder with a movable piston and a force on the piston. This results in pressure in the water. Figure 2-1 shows a piston with an area [A] applying a force [F] to the water. The result is pressure expressed as force per unit area, or $P = F \div A$.

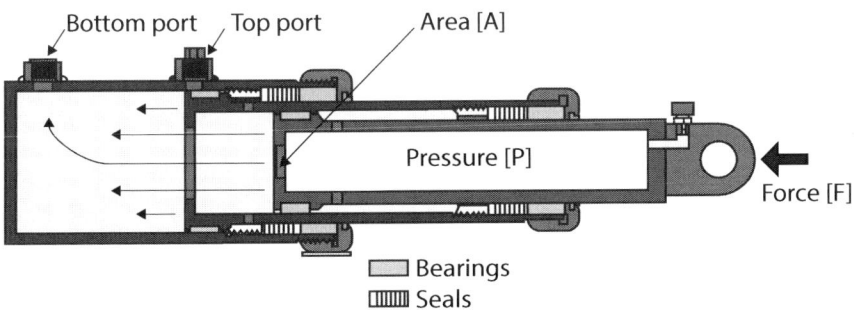

**Figure 2-1**
*Piston applying force to water*

Kinetic energy is energy due to motion, and in regard to pumps it is the velocity of the water. Total energy [E] is the sum of potential and kinetic energy of the water, and it is given by Bernoulli's equation.

**Equation 2-1**
*Bernoulli's equation*

$$E = \left(\frac{P}{\gamma}\right) + z + \left(\frac{v^2}{2g}\right)$$

- kinetic energy
- potential energy

where

| | | |
|---|---|---|
| E | = | energy expressed in terms of length {ft} |
| P | = | pressure {lb/in.$^2$} |
| γ | = | specific weight of the fluid {62.4 lb/ft$^3$; 0.433 lb/in.$^2$-ft*} |
| z | = | elevation of the fluid {ft} |
| v | = | velocity of the fluid {ft/s} |
| g | = | acceleration due to gravity {32.2 ft/s$^2$} |

* To get the units correct, use as 0.433 lb/in.$^2$-ft}

Under ideal conditions energy is conserved, so the velocity could be turned into either pressure or elevation — or both. Likewise, pressure or elevation could be turned into velocity. Because the assumption is that energy is conserved, energy at location 1 is equal to energy at location 2. Refer to figure 2-3. The Bernoulli equation is then written as the following.

**Equation 2-2**
*Bernoulli's equation for two locations*

$$\frac{P_1}{\gamma} + z_1 + \frac{v_1^2}{2g} = \frac{P_2}{\gamma} + z_2 + \frac{v_2^2}{2g}$$

Centrifugal pumps work by first imparting velocity [$v_1$] to the water and then converting it to pressure [$P_2$]. The pump is designed to facilitate this conversion.

## Pressure and Head

Pressure and head differ by the specific weight of the fluid. As discussed, pressure is force per unit area. Force due to a column of water is a result of the weight of the water in the column. The specific weight of the fluid affects the total weight and, hence, affects the pressure. For example, mercury has a specific weight of 845 pounds per cubic feet, so a column of mercury the same height as a column of water has 13.6 times as much pressure as the column of water. Taking the example further, atmospheric pressure is 29.92 inches of mercury or 33.9 feet (406.8 inches) of water (406.8/29.92 = 13.6). One of the reasons mercury is used in barometers is that the height of the column is manageable. It would be very difficult to manage a column of water 33.9 feet high when attempting to use it as an instrument.

An important reason for using head in pump calculations is that pumps actually produce head — not pressure — so the pressure produced by a pump depends on the specific weight of the liquid. Since almost all applications of pumps in irrigation involve water, the standard conversion (for water) from head to pressure is used (see fig. 2-2).

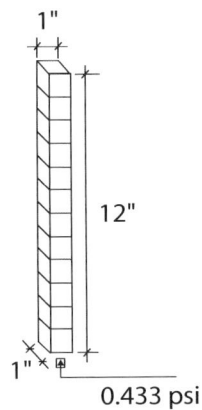

**Figure 2-2**
*Relationship between head and pressure*

The specific weight of water is 62.4 pounds per cubic feet. There are 1,728 cubic inches per cubic feet, so the specific weight of water is 0.0361 pounds per cubic inch.

This column of water 1 foot (12 inches) high is 1 inch by 1 inch in area. The weight in the column is

$$12 \text{ in.} \times 1 \text{ in.}^2 \times 0.0361 \text{ lb/in.}^3 = 0.433 \text{ lb/in.}^2 = 0.433 \text{ psi.}$$

Since the area is 1 square inch, the pressure is 0.433 psi. This is an important conversion to remember. One foot of head (of water) is 0.433 psi. Similarly, 1 psi of pressure is 2.31 feet of head (of water).

## Energy Change from Kinetic to Potential and Potential to Kinetic

The Bernoulli equation shows how energy can change from kinetic to potential and back. Not all conversions are 100 percent efficient, so some energy is lost in the conversion. But, the conversion is essential to pump operation. It also may help later to understand the shape of pump curves. Pump curves demonstrate how the operating conditions dictate how the pump operates and what will be the resulting head and flow.

# Continuity

Continuity is the word used to describe when there is no loss of water. In a water conveyance system, it is assumed that there are no leaks other than designed points of discharge such as emitters, sprinklers, or relief valves. In other words, mass (in this case, water) is conserved. Because water is incompressible in the range of pressure in irrigation systems, conservation of mass implies conservation of volume. Similarly, with no leaks, flow is conserved.

It is common in irrigation systems to have changes in the cross-sectional area of the flow path. The simplest of these is a change in pipe diameter. The example below illustrates continuity in a change of pipe size. The flow [Q] passes through both sections (locations). The flow rate is the same in both sections and is given by equation 2-3.

**Equation 2-3**
*Continuity equation*

$$Q = A \times V$$

where

$Q$ = flow $\{L^3/\text{time}\}$
$A$ = area $\{L^2\}$
$V$ = velocity $\{L/\text{time}\}$

If units of length are feet and time is seconds, flow is in cubic feet per second. If flow is in gallons per minute, velocity is feet per second, and diameter is inches, a conversion factor of 2.45 is needed, so flow is calculated as follows.

**Equation 2-4**
*Pipe flow equation*

$$Q = 2.45 \times V \times D^2$$

where

$Q$ = flow $\{\text{gpm}\}$
$V$ = velocity $\{\text{ft/s}\}$
$D$ = diameter $\{\text{in.}\}$

**Figure 2-3**
*Same flow in two different locations*

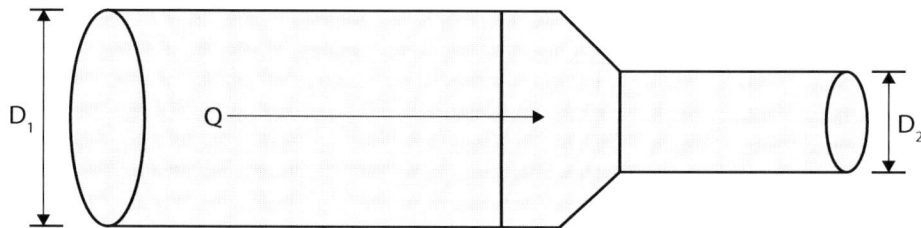

From equation 2-3,

$$Q_1 = A_1 \times V_1 \text{ and } Q_2 = A_2 \times V_2$$

With the same flow in both sections,

$$A_1 \times V_1 = A_2 \times V_2$$

So, if the area goes down, the velocity must go up. If the area goes up, the velocity must go down. The fact that both the energy equation and the continuity equation hold in pumps allows the velocity of the water imparted by the impeller to be converted to head (pressure). While it is general practice to hold velocity in pipes to less than 5 feet per second, velocities in the volute of a centrifugal pump can be as high as 200 feet per second. The volute is designed to withstand these velocities and convert most of the kinetic energy to potential energy. The pump produces velocity in the water pumped, and the shape of the volute with increasing area allows the velocity to be converted to head (potential energy).

# Practice Questions

1. What two types of energy are found in water?

    (1.) _____

    (2.) _____

2. What is the multiplier to convert feet of head to pounds per square inch?

    _____

    _____

    _____

3. What is the continuity equation?

*Chapter*

# 3

# Types of Pumps

## Learning Objectives

The following objectives are the focus of chapter 3:
- understand centrifugal pumps
- understand vertical turbine pumps
- become acquainted with submersible pumps
- become acquainted with jet pumps
- understand positive displacement pumps

## Types of Pumps Used in Irrigation

While many different types of pumps exist, irrigation systems generally employ a few categories including centrifugal, vertical turbine, jet, and positive displacement. Of these, jet pumps and positive displacement pumps are used for a few specific irrigation applications. Most irrigation systems use some variation of either a centrifugal pump or a vertical turbine pump.

A vertical turbine is a specific type of centrifugal pump. Both centrifugal and turbine pumps rely on centrifugal force to add energy to water. Centrifugal force can be observed in many everyday events. For example, after driving a truck out of a muddy field, as the wheels speed up, the mud on the tires is flung radially to the outside because of centrifugal force.

## Centrifugal Pumps

In a centrifugal pump, the volute traps the water that is being forced to the outside of the impeller, and this centrifugal force creates velocity which is converted to potential energy (head) (see fig. 3-1). The centrifugal force developed is proportional to the speed of the rim of the impeller. The faster the outside tip of the impeller rotates, the more head there is to be transferred to the water. The speed of the tip of the impeller can be increased either by increasing the rate of rotation (i.e., revolutions per minute, rpm) or by increasing the diameter of the impeller.

**Figure 3-1**
*Centrifugal pump cross section*

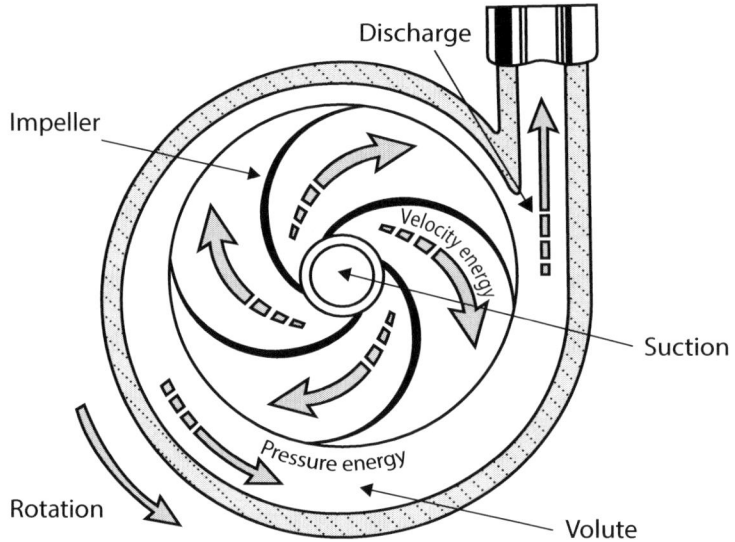

The cross-sectional area between the impellers determines the flow rate that a particular pump can produce. The eye of the impeller is where water enters, and the suction characteristics of any particular pump are dictated by the design of the eye of the impeller.

Centrifugal pumps can also be characterized by how the water flows through them (axial or radial) and the type of impeller in the pump (open, semi-open, or closed). Examples are shown in figure 3-2.

With closed impellers, as shown in figures 3-2 and 3-3, water flows completely within the impeller between the solid top and bottom. A semi-open impeller has the vanes of the impeller exposed on one side. With an open impeller (axial flow), the vanes are open on both sides.

**Figure 3-2**
*Centrifugal pumps and impeller types*

**Figure 3-3**
*Closed impeller*

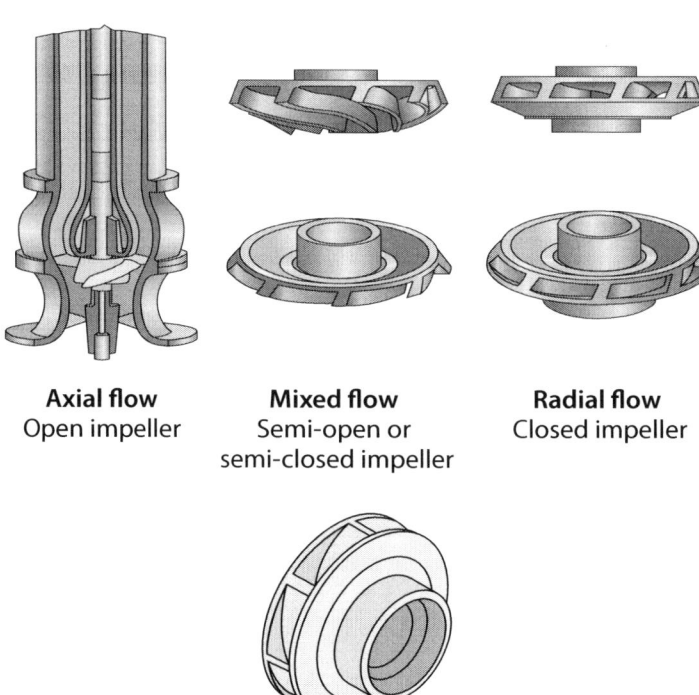

Unlike positive displacement pumps, the flow produced by a centrifugal pump does not remain constant over a range of operating heads. The flow rate is a function of the head that the pump is working against. Conversely, the head produced by a pump is a function of the flow rate through the pump. The flow of a centrifugal pump varies with the operating head. The fact that a pump is producing flow is not an indication that it is working efficiently. The flow that a pump will produce in any particular situation can be predicted from the pump curve and the irrigation system curve, which are discussed later in chapter 8. The head (or pressure) added to the water by a pump approximately equals the difference between the inlet pressure and outlet pressure of the pump (ignoring changes in velocity head). If water is entering a pump at zero pressure, the head that the pump produces will approximately equal the discharge pressure. If the pressure entering the pump is less than zero (a partial vacuum as in the case of a pump mounted above the water level), the discharge pressure will be slightly less than the total head the pump is producing. If the pressure entering the pump is more than zero (e.g., booster pumps, flooded inlets), the discharge pressure will be more than the total head produced by the pump.

## End Suction Horizontal Centrifugal Pumps

End suction horizontal centrifugal pumps are commonly used in irrigation applications. These pumps are generally reasonably priced for irrigation applications. Recent developments in pump design have resulted in improved efficiencies and durability for end suction pumps.

On an end suction horizontal pump, the suction fitting is on the end of the pump, and water enters the suction fitting and goes directly into the eye of the impeller (see fig. 3-4). From the eye, the water is forced radially outward in the volute of the pump, which channels the water to the discharge fitting. The discharge fitting is mounted at right angles to the suction fitting. The discharge is also at right angles to the shaft that is rotating the impeller.

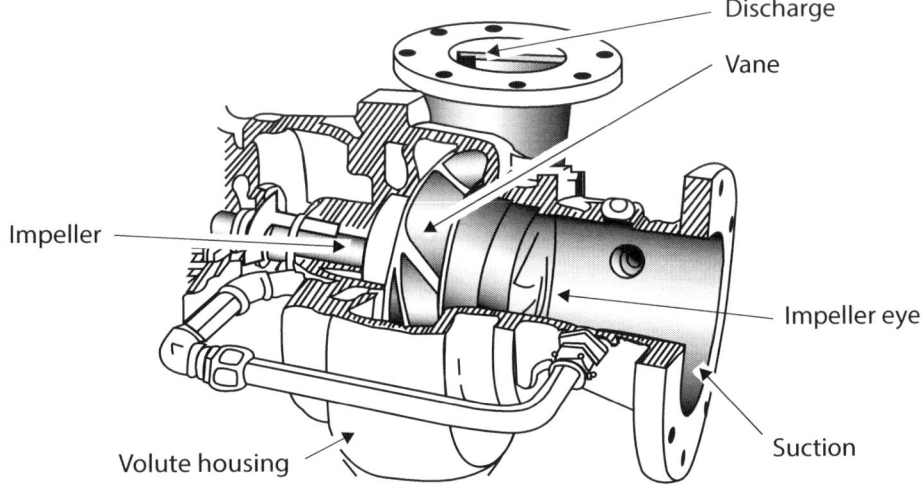

**Figure 3-4**
*End suction centrifugal pump*

With most end suction centrifugal pumps, the volute can be rotated so that the discharge fitting can be mounted in various radial orientations around the suction fitting. The suction fitting is normally one size larger than the discharge fitting, and the nominal size of the pump refers to the pipe size (flanged or threaded) of the discharge fitting. That is, a 6-inch pump would have a standard 6-inch pipe flange as a discharge fitting. The nominal size of a pump is *not* a good indicator of the performance characteristics of the pump.

One disadvantage of end suction centrifugal pumps is that the volute case and the suction piping must be primed before the pump will operate. Priming involves filling the entire volute case and all the suction piping with water. No air may remain in this piping or volute case for priming to be successful. In many irrigation situations, the pump is located higher than the water source. Such an arrangement makes priming more complicated than simply opening a valve. In some situations, priming can be difficult and bothersome, especially if the pump station has not been set up properly.

In special cases, an end suction pump can be mounted vertically with the suction flange down (or even in the water) and a vertical motor mounted on top to produce a vertical end suction pump. Some models use a small propeller mounted in the water to lift water up to the eye of the impeller. Priming is not required with these pumps.

## Split Case, Double Suction Horizontal Pumps

The suction fitting and the discharge fitting of split case, double suction centrifugal pumps are in line with each other (see fig. 3-5). As the water enters the suction fitting it splits into two, approximately equal, streams. Each stream enters the eye of the impeller from opposite sides. From there, as with the end suction pump, the fluid is forced radially outward in the volute of the pump, which channels the water to the discharge fitting mounted on the opposite side of the pump. With split case pumps the suction fitting is often one size larger than the discharge fitting. Again, the nominal size of the pump refers to the pipe size of the discharge fitting. In some cases, split case, double suction pumps contain multiple impellers operating in series.

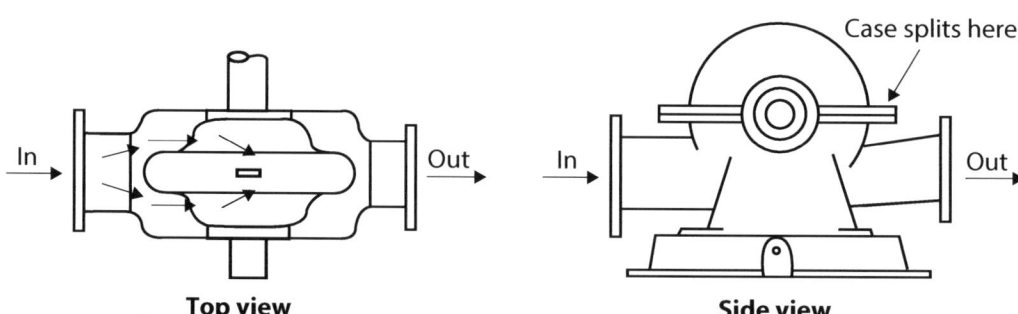

**Figure 3-5**
*Split case, double suction pump*

Split case pumps require priming if they are mounted above the water level. In this case, however, because the discharge fitting is not at the highest point on the pump, priming must often occur at more than one location. Generally, priming a split case pump is more difficult than priming an end suction pump.

Split case pumps are usually more expensive than end suction pumps but usually have better suction characteristics. Split case pumps are well suited as booster pumps or in industrial applications because of straight-through design (requiring less complicated piping) and ease of service. The top half of the pump case can be removed, allowing access to the impeller, shaft, and bearings without disconnecting the pump from the piping. The arrangement makes servicing much easier. In rare cases a horizontal, multistage, split case, centrifugal pump may be used in high-pressure applications, but these are more often found in oil field or industrial applications.

## Vertical Turbine Pumps

In the irrigation industry, what is commonly referred to as a turbine pump is not a true turbine pump. A true turbine (e.g., jet turbine) is made up of a large number of fins, and the pumped fluid (water or air) moves straight through the fins in an axial direction (parallel to the rotating shaft). The turbine pump used for irrigation is actually a mixed flow, multistage, vertical centrifugal pump.

The flow from a vertical turbine pump occurs from the rotation of an impeller, as in a centrifugal pump. The impeller rotates within a bowl assembly (see fig. 3-6). It is mixed flow because the water does not move exactly radially from the impeller, as in the case of horizontal centrifugal pumps, nor does it move exactly axially as in the case of low lift propeller pumps. The water leaves the impeller at an angle to the drive shaft.

**Figure 3-6**
*Vertical turbine pump bowls and impellers*

In many situations, a single stage pump cannot produce enough head to meet the system requirements. Vertical turbine pumps are almost always multistage, employing a series of impellers all mounted on a single drive shaft. The shaft usually extends up to where the power unit is located and the water is discharged. As water leaves the first impeller, paths in the bowl of the pump direct the water to the eye of a second impeller. This process is continued for all impellers (or stages) present in a particular pump.

A vertical turbine pump is simply a number of centrifugal pumps in series. Each successive impeller acts as a booster pump. Each stage adds pressure to the water. A sufficient number of stages (i.e., bowl assemblies) are connected to produce the desired pressure. By using a large number of stages, very high pressures can be produced.

The term "vertical pump" is used because the general direction of water movement is vertically upward through the various stages. At the top of the pump, the water (moving vertically) is channeled through a discharge head into horizontal piping, which supplies the irrigation system. The discharge head also serves as the mounting place for either the electric motor or the gear drive that supplies power to the pump. In some cases, the driving motor is attached below the impellers at the bottom of the well. These are submersible pumps and are described later in this chapter.

Vertical turbine pumps are suited to deep well applications or applications where high pressures are required. Vertical turbine pumps are also well suited to pump sites from rivers or lakes where the power unit must be installed at a point significantly higher than the water source, beyond the suction lift of a typical centrifugal pump. Short shaft vertical turbine pumps are well suited for wet well applications, even where a centrifugal pump could be used, because they do not require priming and therefore are more easily automated.

The distance from the top bowl to the discharge head of a vertical turbine pump can be any length necessary. This length is termed the column length. When pumping from surface water (river, lake, etc.), this distance is usually small, 50 feet or less. In situations where pumping is done from a deep well, the column length can be several hundred feet.

Because at least the first bowl and impeller assembly are installed below the water level, vertical turbine pumps do not require priming. This feature is one of their chief advantages and also makes them easier to include as part of an automated system. Examples would be automatic restarting (in the event of power failure) or remote starting from a pivot point or a centralized control system. One disadvantage of a vertical turbine pump, when pumping from surface water, is that the pump must be mounted on a solid structure. These structures may become complicated and costly in certain situations.

With vertical turbine pumps, the nominal size of the pump refers to the size of the well casing inside of which the pump will fit. A 10-inch vertical turbine pump fits in a 10-inch well casing. The discharge fitting is determined by the size of the discharge head and could be a variety of sizes for any particular pump.

Vertical turbine pumps are generally custom built. Unlike other pumps, a vertical turbine pump is selected as a number of individual components (see fig. 3-7) that are assembled to make an acceptable unit. The components include the following:

- *suction fittings* — suction adapter, bell entrance, suction screen, suction strainer, and/or length of suction pipe
- *bowl assemblies* — including an impeller and the bowl itself
- *discharge adapter* — to adapt from the last bowl assembly to the particular column type and size
- *column pipe and drive line shaft* — assembled in the required length as determined by the elevation of the pump bowls and the elevation of the discharge head (With oil lubrication, the drive shaft is enclosed within a surrounding pipe, which guides oil to the proper parts. Water-lubricated pumps use the pumped water to lubricate the column bearings/bushings.)
- *discharge head* — the base from which the pump assembly is suspended and the component that changes the water flow direction from a vertical to horizontal direction (It also acts as the base to which the electric motor or gear drive is mounted.)
- *power unit* — consists of either a vertical hollow-shaft electric motor or an internal combustion engine coupled to a right-angle gear drive

There are several options regarding these components, especially for the bowl assemblies, impellers, line shaft, and power units. Power units are discussed in more detail in chapter 9.

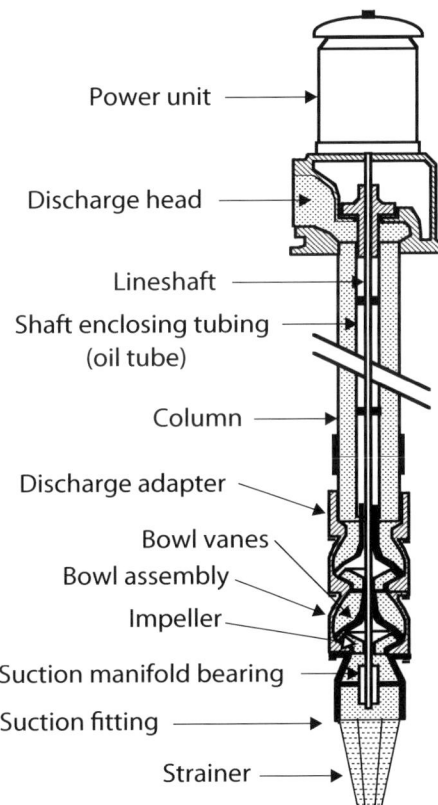

**Figure 3-7**
*Vertical turbine pump components*

## Bowl Materials

A variety of materials are used to construct the bowls, as well as to line the inside of the bowls. In some cases, bowls are lined with material that is resistant to corrosion by the fluid being pumped. In cases where energy efficiency is critical, a smooth coating may be applied to the bowls to improve the efficiency.

## Closed and Semi-Open Impellers

Closed or enclosed impellers (see figs. 3-3, 3-6, and 3-8) have a solid top and bottom that make the water flow completely within the impeller. A semi-open impeller (see figs. 3-2 and 3-8) has the vanes of the impeller exposed on one side.

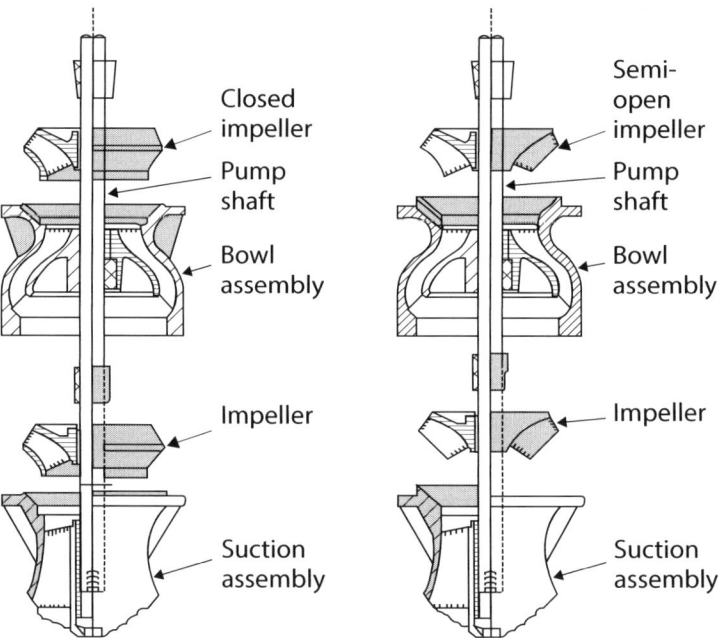

**Figure 3-8**
*Vertical turbine impeller types*

The performance of closed impellers is dictated by the size of the impeller and the clearance between the outside of the impeller and the inside diameter of the bowls. Because all of the flow is within the impeller itself, the vertical clearance between the impeller and the bowl (raising and lowering the impeller) does not affect the pump flow/pressure relationships.

The performance of semi-open impellers can be adjusted at the top of the motor (or gear drive) by slightly raising or lowering the line shaft and the impellers, thereby changing the clearance between the vanes of each impeller and the bottom of each bowl. This can be an advantage because the performance of the pump can be modified. However, if this adjustment is not made correctly, the performance of the pump may be much different than that shown on the manufacturer's pump curves.

## Oil Lube and Product (Water) Lube Line Shafts

Most vertical turbine pumps used with irrigation systems use the pumped fluid (water) to lubricate the bearings that surround the line shaft from the bowls to the discharge head. This option is the most economical lubrication method and well suited to irrigation as long as the water is nearly free of sediment. This option also provides a slight hydraulic advantage because the space taken up in the middle of the column by the line shaft is smaller and there is more area for the water to flow. Therefore, the friction losses within the column are less.

If the water being pumped contains a significant amount of sand or other impurities that could damage the bearings, it is better to use an oil-lubricated line shaft. In this case, a small diameter pipe (oil tube) surrounds the line shaft and bearings for the full length of the line shaft. Oil is dripped into this oil tube from the top and flows down to lubricate the bearings. Care must be taken to ensure that the oil used to lubricate the bearings does not pollute the water source (groundwater or river).

## Line Shaft Size

Once the impellers and number of bowls have been selected and the brake horsepower [Bhp] requirements are calculated, the designer needs to determine what size line shaft to use. The line shaft must be large enough to transmit the total brake horsepower that is being transferred to the impellers. However, too large a line shaft results in an unnecessarily expensive pumping unit. Also, the friction losses in the column increase as the size of the shaft increases. The increase in friction loss is significant only for turbine pumps installed in deep well applications where the column is hundreds of feet long.

Another consideration of deep well applications is the line shaft stretch. When the pump is not operating, there is practically no difference in pressure between the top of the top impeller and the bottom of the bottom impeller. However, once the pump starts, the total pressure produced by the pump pushes down on the top impeller, and this pressure exerts a downward force that causes the line shaft to stretch. If the line shaft stretches an excessive amount, the impellers may bottom out on the bottom of the bowls, and the line shaft and impellers will not rotate. Conversely, if the line shaft is adjusted so it rotates freely under pressure, when the pump turns off, the line shaft shortens in length and the impellers may bind on the top of the bowls. To prevent excessive line shaft stretch, a larger diameter line shaft must be used. Vertical turbine pump manufacturers provide tables and sample calculations to aid designers in selecting the correct line shaft diameter.

# Submersible Pumps

Although many pumps are submersible (i.e., the pump is in the water), the term "submersible pump" typically refers to vertical turbine pumps with the impellers, bowls, and the electric motor powering the pump all installed below the water level. The bowls and impellers used in submersible irrigation pumps are like (sometimes identical to) those used in vertical turbine pumps. Submersible pumps are also often used in wells.

The most common submersible example for irrigation systems is a vertical multistage turbine pump with the motor installed in the well casing or sump below the pump bowls (see fig. 3-9). It is connected to the impellers with a short shaft. Because the motor and pump bowls are one assembly, they are mounted together and alignment and line-shaft lubrication are not a problem. The column (discharge piping, pipe drop) is not required to be perfectly straight as there is no line shaft in the pump column. The full column pipe area is available for water flow. The wiring to the motor is attached to the exterior of this pipe with the motor controls normally above grade. The top of the pump discharge is simply connected to a discharge pipe that runs to the top of the well or sump.

**Figure 3-9**
*Submersible pump components*

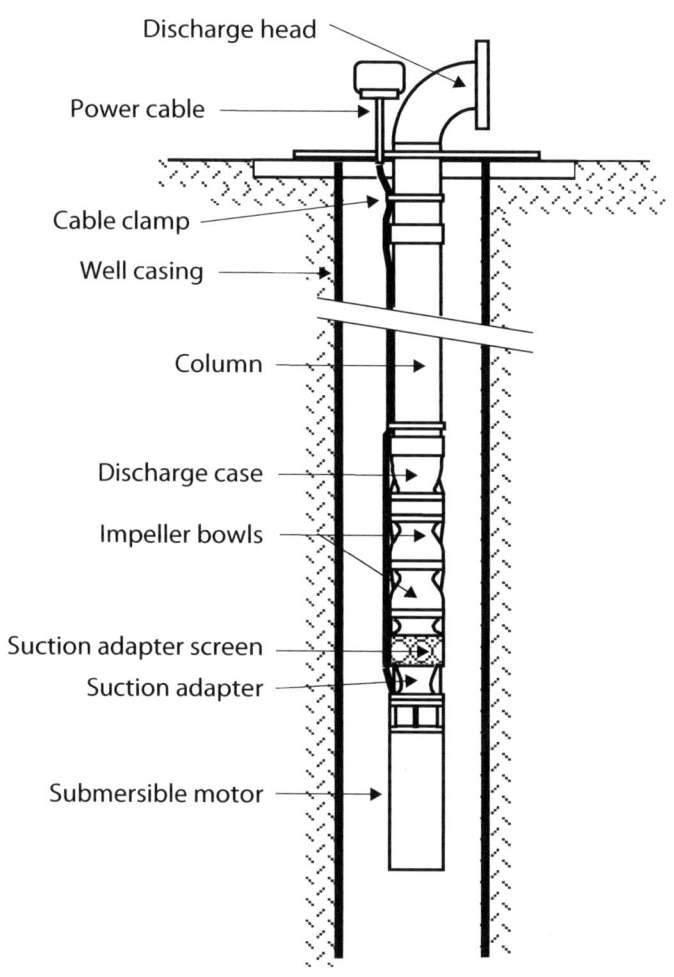

Submersible pumps require special motors because they often must be small to fit in well casings, and they are specially constructed to operate when submersed in water. Often the outside diameter of the motor is the same as the diameter of the bowls of the pump.

It is very important that water flow past the motor before entering the pump so that proper motor cooling occurs. Therefore, if a submersible pump is used in an open water application or if water enters at points above the pump motor, some kind of outer pipe (inducer sleeve) is needed to ensure that the water flows up past the motor before entering the bowls (see fig. 3-10).

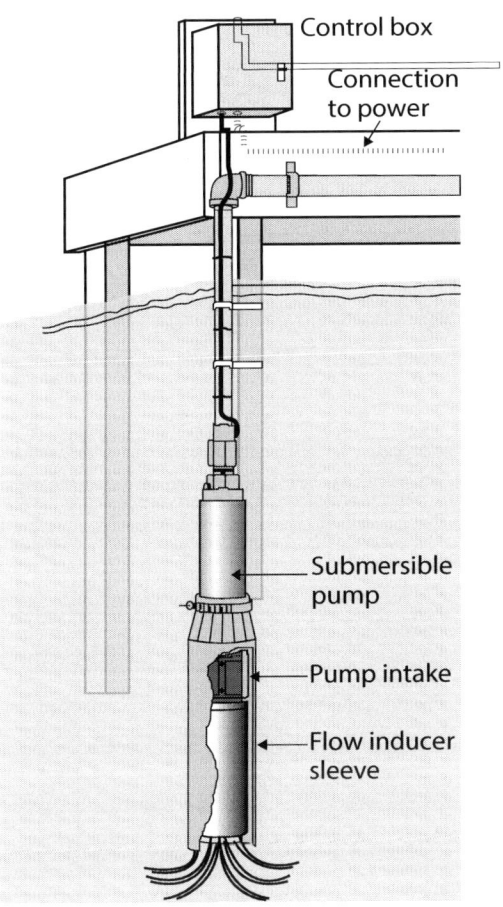

**Figure 3-10**
*Inducer sleeve*

Submersible pumps are usually more expensive than standard vertical turbine pumps because of the cost of the submersible motor. For irrigation applications they are often used in specialized applications (e.g., urban or golf course). Because the motor is below the surface, operation is quieter and vandalism is less of a problem than with surface-mounted power units. Submersible pumps have also been used successfully in an inclined position as part of a river pumping application.

# Jet Pumps

In some cases, jet pumps are used in irrigation applications. They are often used in shallow wells and can even be used in deep wells. These pumps appear to lift water a great distance, but they do not actually lift water. They are a type of centrifugal pump that discharges some of the water pumped into the irrigation system. Some of the water is recirculated back down the well where it is forced through a nozzle in the well water; the water flowing from the nozzle is forced back to the surface and brings water from the well along with it. Their use is best suited to small flow rate systems where pump efficiency and energy costs are not crucial considerations (e.g., residential or small turf irrigation).

# Positive Displacement Pumps

There are numerous designs of positive displacement pumps. A very important feature of positive displacement pumps is that they deliver essentially the same flow rate (at a constant rotation speed) regardless of the pressure requirement. For that reason, they are often used to inject chemicals such as herbicides, pesticides, and fertilizers into the irrigation water knowing that the flow rate will not change.

## Piston Pump

A positive displacement pump creates flow by entrapping a volume of water in some sort of chamber and forcing that volume to move. The most common of positive displacement pumps is the piston pump.

## Peristaltic Pump

Since the development of flexible tubing capable of withstanding millions of cycles of flex, the peristaltic pump has become the positive displacement pump of choice. As the name implies, the peristaltic pump mimics the peristalsis action of the human digestive system that moves food through the intestines. Figure 3-11 is an example of a peristaltic pump. As this three-lobe pump rotates, each lobe with a roller traps a fixed volume of liquid (darker section) and moves it around the flexible tube to the exit (lighter section). Where the roller is in contact with the tube, it is pinched completely flat so that no liquid can pass. For a given rotation speed, the same amount of liquid is transferred through the pump. If the exit pressure becomes too high, the pump power unit may slow down or stall.

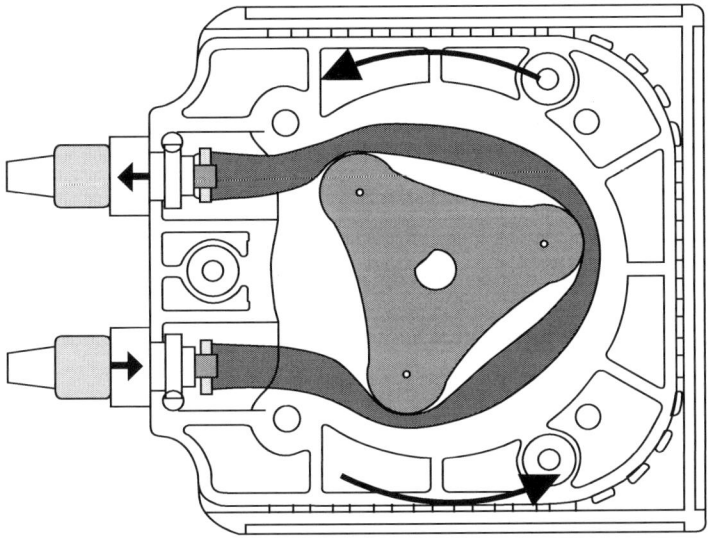

**Figure 3-11**
*Peristaltic Pump*

An advantage of the tube peristaltic pump is that none of the liquid being pumped comes in contact with the rotating parts. This prevents contamination and/or corrosion of the moving parts.

## Other Positive Displacement Pumps

There are numerous other types of positive displacement pumps. Gear pumps are commonly used in oil hydraulic systems and lobe pumps are commonly used in air compressors. Diaphragm pumps are used for priming centrifugal pumps and for fuel pumps in automobiles. The rope pump is an intriguing design that has been used for centuries (see fig. 3-12). It uses a rope on a pulley system with the lift side enclosed in a pipe. The rope pump has knots in the rope so that it somewhat seals within the pipe, and the volume pumped is contained between the knots. According to one source, more than 4 million people worldwide use rope pumps for domestic and irrigation water.

**Figure 3-12**
*Rope pump*

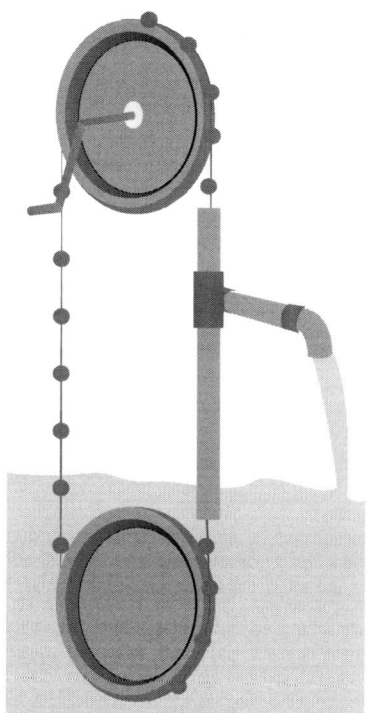

# Practice Questions

1. Name four types of pumps.

    (1.) _____

    (2.) _____

    (3.) _____

    (4.) _____

2. Which type of pump is most likely used in deep wells?

    _____

3. Which type of pump has the motor in the well casing below the water level?

    _____

4. Which type of pump would likely be used for chemical injection?

    _____

# Chapter 4

# Basics of Pumps: Force, Work, and Power

## Learning Objectives

The following objectives are the focus of chapter 4:
- understand pump work
- understand pump power
- calculate pump power and understand the effect of efficiency

## Work

All pumps take some form of external energy and use it to produce pressure — perhaps in the form of lift and flow. The amount of energy (work) is defined as a force through a distance. For example, in the case of a simple bucket on a rope drawing water from a well, the force is the weight of the water in the bucket. It has already been shown that head of water represents the weight of the water. The distance is the height to which the bucket is drawn.

## Power

Power is the rate at which work is done, or work per unit of time. It could be thought of as head multiplied by the distance over which the water is pumped divided by the time it took to pump it. Velocity is distance [d] divided by time [t], or d/t. Look at head multiplied by flow in the equation 4-1 progression below showing force [F], work [W], and power [P].

$$H \times Q = H \times A \times V$$

But: $V = d/t$, so

$$H \times Q = \underbrace{H \times A}_{} \times d/t$$

*Head (pressure) multiplied by area is force [F].*

$$\underbrace{F \times d}_{} / t$$

*Force multiplied by distance is work [W].*

$$\underbrace{W/t}_{}$$

*Work per unit of time is power [P].*

$$P$$

**Equation 4-1**
*Pump power from head and flow*

Chapter 4: Basics of Pumps: Force, Work, and Power

**Equation 4-2**
*Water horsepower*

So, H × Q represents power. The units of this relationship are not consistent with units normally used in irrigation. Water power is generally written as horsepower, so the H × Q relationship with appropriate units is given as equation 4-2.

$$\text{Whp } \{hp\} = \frac{H \{ft\} \times Q \{gpm\}}{3{,}960}$$

## Pump Efficiency

**Equation 4-3**
*Brake horsepower*

Equation 4-2 represents the power produced by the pump, but it does not account for the efficiency of the pump. The power required at the drive shaft is called brake horsepower and is given by equation 4-3.

$$\text{Bhp} = \frac{\text{Whp}}{E_{pump}} = \frac{H \{ft\} \times Q \{gpm\}}{3{,}960 \times E_{pump}}$$

where
- Whp = water horsepower {hp}
- $E_{pump}$ = pump efficiency converted to a decimal (e.g., 76% = 0.76)
- H = total dynamic head {ft}
- Q = flow rate {gpm}

# Practice Questions

1. What is the water horsepower equation?

2. What is the brake horsepower equation?

3. How much brake horsepower would it take to pump 550 gpm at 150 feet of head with a pump efficiency of 75 percent?

   _____ Bhp

Chapter

# 5

# Pump Performance

## Learning Objectives

The following objectives are the focus of chapter 5:
- determine design flow rate
- determine total dynamic head
- understand net positive suction head
- understand cavitation and its causes
- introduce pump curves and how they can be used to determine if a pump will perform effectively

## Selection Criteria

In many cases, selecting a pump is a straightforward process. Once the designer knows the flow and pressure required, pump curves can be examined and a pump can be selected that efficiently produces the flow and pressure required. Usually it is slightly more complicated, and a basic knowledge of pump curves will aid greatly in selecting the best pump for any particular application.

To adequately do the required job, a pump must be able to deliver the required flow at the correct head (or pressure). The range of head and flows a particular pump can produce is shown graphically on charts known as pump curves.

A basic, simplified pump curve is shown in figure 5-1. The two thicker curved lines sloping from the left down to the right are the basic flow [Q] vs. head [H] curves and show the head/pressure a pump can produce at different flow rates for two different impeller diameters. This set of curves show that as flow increases, head decreases. At very low flows the pump produces its maximum head. When the flow rate is zero (pumping against a closed valve), the pump produces its "shutoff head." For this simplified case, only two Q-H curves are shown (for 11-inch and 12-inch impellers). Actual pump curves show data for different impeller diameters (or different pump impeller speeds or revolutions per minute). Also shown on the pump curves is the pump efficiency at different operating points. A pump should be selected that will be operating at points within its high efficiency range most of the time. For the pump shown in figure 5-1, this would be in the 1,200–1,700 gpm and 80–120 feet total dynamic head operating range.

**Figure 5-1**
*Simple pump curve*

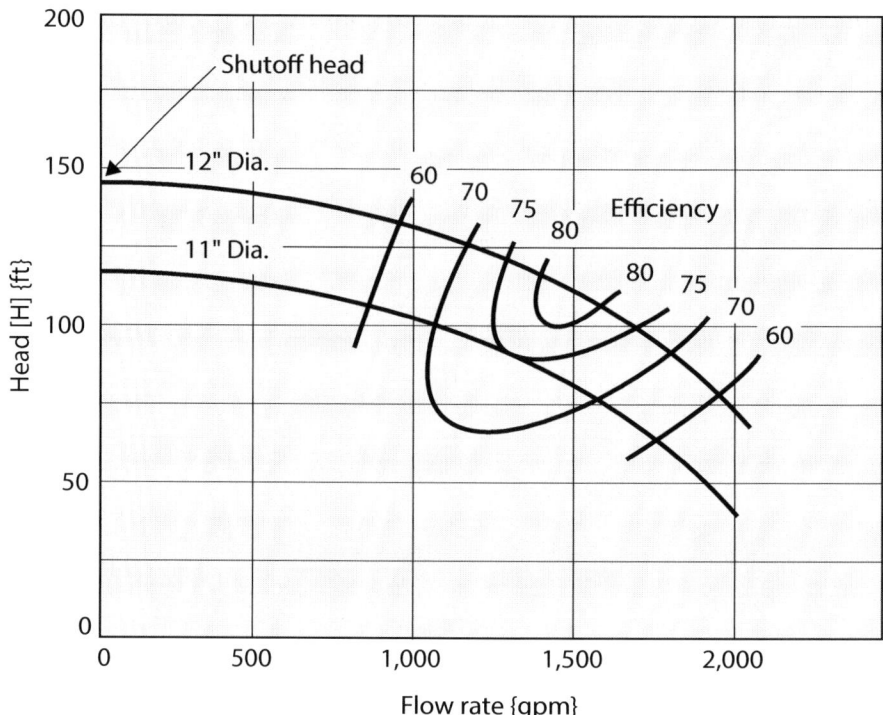

To select a pump, it is necessary to have a basic understanding of hydraulics and the conversion of energy forms as was presented in chapter 2.

# Design Flow Rates

Determining the design flow rate for a pump system normally is straightforward. The simplest case is with a pressurized system with a fixed number of discharge points such as a sprinkler or emitter system. However, it is common practice for the system to have zones with different flows or flow conditions resulting in different design flows. Designers try to keep the flow in different zones about the same, but it is likely that there will be a range of flows required. The system capacity requirement (flow rate) based upon the area to be irrigated, evapotranspiration, and other factors is given by equation 5-1 below.

**Equation 5-1**
*System capacity requirement*

$$Q = \frac{\left[(ET_o \times K_L) - R_E\right] \times A \times 0.623}{E_a}$$

where

$Q$ = minimum system capacity required {gal/day}
$ET_o$ = reference evapotranspiration {in./day}
$K_L$ = landscape coefficient
$R_E$ = effective rainfall {in./day}
$A$ = area to be irrigated {ft²}
0.623 = conversion factor for area and inches into gallons {gal/in. × ft²}
$E_a$ = application efficiency (decimal equivalent, suggest 0.80)

This is the minimum design flow rate. It is further amended by the amount of time the system can be operated each day. For example, if the system can only be operated 50 percent of the time, the minimum capacity must be twice as large as calculated by equation 5-1.

**Example 5-1**
*System capacity*

Assume that the system can be operated 18 hours per day to irrigate 10 acres, that the reference ET for the design period is 0.25 in./day, that the landscape coefficient is 0.8, and the application efficiency is 0.80. Determine the system capacity for the following conditions. For the design period, there is no effective rainfall.

$$Q = \frac{\left[(ET_O \times K_L) - R_E\right] \times A \times 0.623}{E_a}$$

$$= \frac{\left[\left(0.25\left\{\frac{in.}{day}\right\} \times 0.80\right) - 0.0\right] \times 10\{A\} \times 43{,}560\left\{\frac{ft^2}{A}\right\} \times 0.623\left\{\frac{gal}{in. \times ft^2}\right\}}{0.80}$$

$$= \frac{(0.20 \times 10 \times 43{,}560 \times 0.623)}{0.80}\left\{\frac{gal}{day}\right\} = 67{,}845\left\{\frac{gal}{day}\right\}$$

The average, maximum, and minimum flow rates sometimes vary over a large range. The flow rates used to select a pumping system must include the maximum expected flow rate for the system, even if only for a small percentage of the time. Variable flow conditions exist with systems such as automated golf course and landscape systems, center pivot systems with corner arms, and set-type systems when the number of sprinklers operating varies for the different zones. Irregular-shaped fields necessitate that the flow change during the irrigation season as different areas or zones are irrigated. Therefore, in addition to the maximum flow rate, the range of operating flow rates must be taken into account.

Considering only the maximum flow rate can result in excessive pressure when the flow is reduced, this additional pressure could cause the following problems:

- Excessive nozzle pressure can lead to misting of sprinkler discharge and low application efficiencies.
- Increased sprinkler flow can result in application depths greater than expected and excessive application rates that exceed the intake rate of the soil, resulting in runoff and soil erosion.
- Increased pressure at lower flow, where there are extreme differences between the minimum and maximum flow rates, may be sufficient to exceed the pressure rating of some of the irrigation system components.
- The lower flow rate may be so far to the left of the peak efficiency range that the load on the power unit may actually increase. (At the extreme ends of all pump curves the efficiency is poor but unknown. The efficiency lines are not necessarily spaced similarly to those shown on the curve.)
- As a general rule, do not operate a pump outside the range where efficiencies are plotted without consulting with the pump manufacturer.

When the flow rate is variable, it is important to remember that the pressure losses within the system decrease as the flow rate decreases, aggravating the problems of excessive pressure. Ideally, the pumping station should produce less pressure when the flow is reduced. This situation is possible only when some kind of automation, variable frequency drive [VFD] controls, and/or multiple pumping units are designed into the pump station.

In most single pump situations, an alternative is to select a pump with a "flat" curve (see fig. 5-2). A flat curve means that the pump produces roughly the same pressure over a wide range of flow rates. Flat pump curves provide flexibility to modify the system in the future without changing the pressure available at the pump station. Conversely, a system with a relatively constant desired flow rate but varying pressure requirements due to large elevation changes (e.g., a center pivot system on extremely rolling land) would be best served by a pumping unit with a "steep" pump curve (see fig. 5-2). A steep curve means that the pump produces roughly the same flow rate even if the pressure/head varies significantly.

**Figure 5-2**
*Flat vs. steep pump curves*

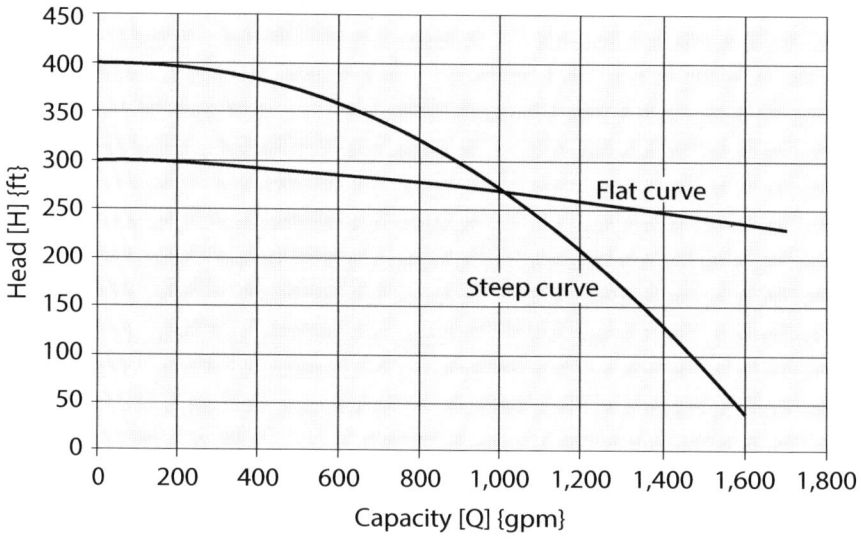

When establishing the flow rate in the pump selection process, it is a good idea to estimate changes to the system that are probable in the future. If the system will be used for filling a pond or some other high flow/low pressure application, this must also be considered. Even though a desired application will be for only a short duration, it is necessary to consider this situation when selecting the pump. In some cases it is impossible for a given pumping unit to provide these incidental flows, so specialized valving or a different pump must be used.

In summary, all possible flow rates must be considered when selecting a pump. The design flow rate may actually be a range of flows. If this is the case, an estimate of the percentage of the time that the pump will operate at each of the flow rates is required. This estimate allows the selection of a pump that performs at reasonable efficiency over the range of expected flow rates.

# Design Pressure — Total Dynamic Head

For any particular irrigation system, there is a total dynamic head [TDH] that is required for the irrigation system to operate correctly. TDH is the pressure necessary to overcome the friction losses in pipes and fittings and to overcome changes in elevation between the water source and the point at which water is discharged and still leave sufficient pressure at the sprinklers or other application devices to allow them to operate correctly.

TDH is the total of several component heads (see fig. 5-3). This section outlines each of these component heads and provides instructions on calculating TDH. Not every component head exists for every pumping situation.

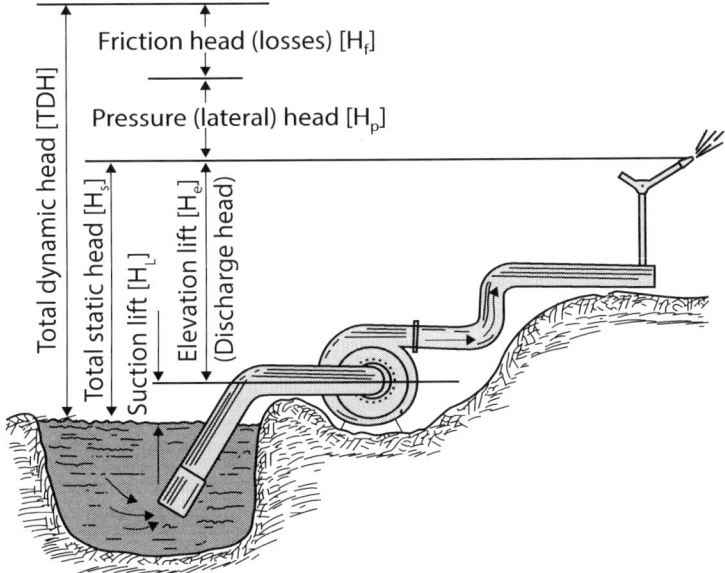

Figure 5-3
Components of TDH

## Suction Lift

When a centrifugal pump is installed above the water source, the pump relies on atmospheric pressure to push down on the surface of the water and force water into the impeller. The pump creates a partial vacuum at the eye of the impeller, and the atmospheric pressure forces the water up into the impeller where it is pressurized and forced out of the discharge of the pump.

In this case the vertical distance from the surface of the water to the eye of the impeller is termed the suction lift [$H_L$] of the pump. Because normal atmospheric pressure (14.7 psi at sea level) is approximately equal to 34 feet of water, it is physically impossible to have a static suction lift greater than 34 feet. In actual practice, lifts in excess of 18 feet are seldom successful.

When a pump is installed as a booster pump or when the level of the water source is above the eye of the impeller (flooded suction), the pressure at the impeller is more than atmospheric. In this case, the suction lift is negative (i.e., positive suction head). The pump can produce less head and consume less energy because some of the desired discharge pressure is already present.

## Elevation Lift or Discharge Head

Irrigation systems transport water to a location where it leaves the system through a pipe or some other application device (sprinklers, emitters, etc.). Normally the point at which water leaves the system is higher than the point at which the pump is installed. This vertical distance, or the difference in elevation between the point at which water leaves the system and the elevation of the impeller, is termed the discharge head or elevation lift [$H_e$]. If the point of discharge is at a lower elevation than the impeller, the elevation lift is negative. In this situation, a pump may still be needed to lift water from the source, overcome friction losses, and supply pressure to operate the system.

## Total Static Head

The suction lift and elevation lift are both termed static heads because they exist even when no water is moving (static conditions). They do not change regardless of the flow rate in the system. This head is the difference in elevation between where the water is and where the water leaves the system. For most systems, the discharge point is above the original water level, and the total static head is equal to the elevation lift (discharge head) plus the suction lift, as given in equation 5-2.

**Equation 5-2**
*Total static head*

$$\text{Total static head } [H_S] = \text{Suction lift } [H_L] + \text{Elevation lift } [H_e]$$

For systems where either the suction lift or elevation lift is negative, the total static head is still the difference between the elevations of the water source and the outlet. As long as a pumping unit can produce something more than the total static head, some water will move through the system.

## Pressure (Lateral) Head

Pressure head [$H_p$] refers to the pressure required at the end nozzle, or other outlet device, on the irrigation system. For large sprinklers this pressure head may be as much as 100 psi. For gravity irrigation systems, where the water flows freely from the end of the pipe, the pressure head would be zero. For systems such as microirrigation or gated pipe, the pressure head is small but is still an important consideration in pump selection.

If the sprinkler flow rate increases, the pressure head required to produce that flow also increases if the nozzle size is not changed. The pressure head must be sufficient to provide the desired flow rate, provide adequate sprinkler stream breakup and also result in a large enough radius of throw so that sufficient wetting pattern overlap is obtained.

## Friction Head (Losses)

Anytime water flows, whether it is in an open channel or a pipeline, energy (pressure) is lost because of the effect of friction between the moving water and the walls of the channel/pipe. These friction losses [$H_f$] increase significantly as the velocity (and hence the flow rate) of the water increases. The pump must be able to overcome these friction losses or there will not be sufficient pressure at the sprinklers or other water outlet devices. Head losses in filters or other operating components such as valves or pressure regulators must be included. In low-pressure microirrigation systems, these commonly called "minor" losses can be a major part of the total friction losses.

## Velocity Head

Most of the energy that a pump adds to water is in the form of velocity, and it is converted to pressure in the volute. The faster an object is moving, the more energy it contains. In fluid systems, the faster the fluid moves, the larger the velocity head. Velocity head [$H_v$], the amount of energy or head {feet} needed to obtain a certain velocity, is given by equation 5-3.

$$H_v = \frac{V^2}{2 \times g}$$

**Equation 5-3**
*Velocity head*

In a pipeline in which the velocity of water flow is 5 feet per second, the velocity head would be calculated as follows:

$$H_v = \frac{5^2}{2 \times 32.2} = \frac{25}{64.4} = 0.39 \text{ ft} = 0.17 \text{ psi}$$

**Example 5-2**
*Calculating velocity head*

This example shows that velocity head in irrigation systems is normally quite small in comparison to other heads. As a result, the velocity head is often ignored in calculations related to irrigation pumping. However, in situations where high velocities and low pressures are required, velocity head can be a significant component of the TDH. Often the fittings close to the discharge of a pump are relatively small, and the water is moving at a high velocity through these fittings. A significant amount of the energy contained in this high velocity water is in the form of velocity head. A pressure gauge mounted perpendicular to the direction of flow (as is normally the case) measures pressure head only, not velocity head. Therefore, it is desirable to mount pressure gauges downstream of all small fittings preferably in a pipe with the same diameter as the pipeline serving the irrigation system.

In some cases velocity head can be a problem where flow paths are small and velocities are high. On the suction side, if the velocity is high and the pressure is low, cavitation can occur. Cavitation is discussed later in this chapter.

## Calculation of Total Dynamic Head

The total amount of pressure that the pump must provide at a desired flow rate is equal to the TDH. Once the TDH is known, pump curves can be consulted and a pump can be selected that produces the necessary TDH at the various design flow rates.

TDH is calculated as the sum of the heads that have been described previously and is shown in equation 5-4.

**Equation 5-4**
*Total dynamic head*

TDH = Suction lift + Elevation lift + Friction head + Pressure head + Velocity head

or

$$TDH = H_L + H_e + H_f + H_p + H_v$$

When calculating TDH, it is imperative that all of the components be in the same units {psi or feet}.

Because most irrigation pumps operate under a number of flow and TDH conditions, it helps in the pump selection process to create a table that lists each condition and includes columns for the following items:

- condition description (such as "zone no.," "corner arm off")
- percentage of the time this condition exists
- design flow rate
- TDH
- pump efficiency
- required brake horsepower
- notes regarding this condition

Each flow rate and TDH condition can then be marked on a copy of the pump curves being considered to ensure that a particular pump will provide the necessary head under all design conditions. A sample of this load profile is shown in table 5-1.

**Table 5-1**
*Pump operating conditions*

| Condition | Time {%} | Design flow rate {gpm} | TDH required {ft} | Pump efficiency {%} | Bhp required {hp} | Notes |
|---|---|---|---|---|---|---|
| High elevation, corner arm off | 35 | 900 | 200 | 80 | 56.8 | North side |
| High elevation, corner arm on | 15 | 1,200 | 220 | 77 | 86.6 | NW/NE corners |
| Low elevation, corner arm off | 35 | 900 | 160 | 78 | 46.6 | South side |
| Low elevation, corner arm on | 15 | 1,200 | 180 | 76 | 71.8 | SW/SE corners |

Table 5-1 shows an example for an agricultural center pivot system. Typically, turf systems (e.g., golf courses) would have even more widely varying flow rates and therefore widely varying power requirements. That is one reason why multiple pump stations or prepackaged pump stations with flow and/or pressure control are more common in turf applications.

# Net Positive Suction Head

Net positive suction head [NPSH] refers to the absolute pressure at the eye of the impeller. If the absolute pressure available is too low, the water will begin to vaporize (i.e., it boils). Cavitation occurs because the vaporizing water forms a great many small bubbles that eventually implode against the impeller surface. The force of these imploding bubbles erodes and damages the impeller. For this reason, it is critical that the net positive suction head available [NPSHa] to the pump exceed the minimum NPSH required for the pump as determined by the manufacturer. Ensuring adequate NPSH requires an understanding of its component factors and the consequences of inadequate pressure.

NPSH is usually significant only when the pump is operating in a suction lift situation, that is, when the pump is lifting water from a free water surface that is lower than the pump. In situations where the water level is above the eye of the impeller, or the water enters the pump under positive pressure, NPSH calculations are not usually required. In these cases, however, submergence may be an issue.

# Submergence

In some cases vortexing can cause air to enter the suction/inlet piping of a pump. Setting the pump inlet at a level with sufficient submergence below the water level will prevent vortexing. The well draw-down level or the lowest level for a fluctuating surface water level must be considered when determining at what level to set the pump inlet to provide the necessary submergence. Turbine pumps have special requirements for required submergence. Each model has different requirements; therefore, the manufacturer's literature must be consulted. NPSH is normally not a problem with impellers located below the water level (e.g., vertical turbine pumps), but submergence can be.

## Atmospheric Pressure

Pumps do not suck water up out of the surface supply. Water is pushed into a centrifugal pump by the pressure of the atmosphere. Under standard atmospheric conditions, atmospheric pressure is 14.7 psi (34 feet of water) or 29.92 inches of mercury.

Atmospheric pressure is a result of the weight of the atmosphere bearing down on the surface of the earth. Where there is less atmosphere bearing down, such as at high elevation, the atmospheric pressure is less. A practical significance of this is that when pumping at higher elevations, less pressure is available to push the water up into the pump, which may make the pumping situation more problematic.

If a perfect vacuum could be created in a pipe, water would rise vertically a distance equal to the atmospheric pressure on the water. If that pressure were 14.7 psi, which is the normal pressure at sea level, water would rise a distance of 34 feet. At higher elevations, water would not rise this far because the atmospheric pressure is less. For example, the normal atmospheric pressure at Denver, Colorado (elevation

5,200 feet), is about 12 psi, which converts to a water column only 28 feet high.

When pumping from a free water surface, the amount of pressure remaining at the eye of the impeller is always less than the surrounding atmospheric pressure. The reason this pressure is less is because of the elevation of the eye above the water surface and friction losses in the suction piping system. The total pressure or head remaining at the eye of the impeller is NPSHa. The NPSHa *must be greater* than the net positive suction head required [NPSHr] for the pump to perform properly.

## Calculating the NPSH Available

Before constructing the pump station, the NPSHa at the eye of the impeller must be calculated to ensure that it is sufficient for the pumps to operate correctly. The NPSHa is equal to the atmospheric pressure [$H_a$] available at the source of the water minus the following:

- vertical distance to the eye of the impeller from the water level in the pond or sump [$H_L$] (The minimum seasonal water level for sources such as rivers, ponds, and canals must be assumed when calculating $H_L$.)
- friction losses such as in the suction fittings, piping, screen box, etc. [$H_f$]
- vapor pressure [$H_{vp}$] of the water being pumped, at the temperature of the water being pumped

or

**Equation 5-5**
*Net positive suction head available*

$$NPSH_a = H_a - H_L - H_f - H_{vp}$$

All of these heads must be in the same units {feet or psi}. These calculations should be done assuming the highest design flow rate identified. Atmospheric pressure, sometimes referred to as barometric pressure, varies from day to day. This daily variance is so small relative to NPSH calculations that it can be ignored. It is safe to use a standard atmospheric pressure based on the elevation at the pump site.

When calculating the NPSHa, the vertical suction lift is relatively straightforward and can be determined by a survey or the use of a hand level. Normal atmospheric pressure can be obtained from a chart if the elevation above sea level is known. The vapor pressure of the water can be determined from a table based on the temperature of the water.

The friction losses in the suction piping are somewhat more complicated to calculate, and every part of the suction piping must be considered. This can include any or all of the following:

- screen box
- foot valve
- entrance losses into the suction pipe
- friction losses in the suction pipe
- friction losses in any fittings (elbows, etc.) in the line between the water source and the pump

Generally, losses are calculated in individual fittings and then totaled to get the total suction line friction losses. Because the total friction loss on the suction side is crucial in determining whether or not a pump cavitates, it is very important to keep the suction piping as simple as possible. Insofar as possible, tees, elbows, valves, and other fittings should be installed on the discharge side of the pump rather than the suction side. These calculations should be done assuming the highest design flow rate identified.

## Measuring the NPSH Available

For existing pump stations, it may be necessary to take a physical measurement of the NPSHa in situations where NPSHa calculations are difficult or when troubleshooting. The NPSHa can be measured using a vacuum gauge, pressure transducer, manometer, or an absolute pressure gauge. The latter has a scale in the area below zero gauge pressure.

NPSHa measurements are advisable even in situations where pump problems are not apparent from calculations or observed by an experienced person. In the experience of many pump testers, the vast majority of pump problems originate on the suction or inlet side of the pump. Using a gauge to take an NPSHa reading and comparing the result to the NPSHr from the pump curve is often the first step in troubleshooting a pump that is not performing properly.

## NPSH Required

The absolute pressure that must be present at the eye of the impeller for any particular pump in order to prevent cavitation is termed the net positive suction head required [NPSHr]. NPSHr varies according to the design of each pump and the head and flow rate. The NPSHr can be read from the pump curve for any particular pump at any particular operating point (see figs. 5-7, 5-8, and 5-9). However, different manufacturers present NPSHr data in different forms. The person selecting a pump must be careful to read the legend of the pump curve.

The design of the pump site and suction piping and the vertical suction lift must be such that NPSHa exceeds NPSHr for all operating conditions. It is important to consider high flow/low head situations as well (e.g., pond filling). Although these conditions may be considered incidental, occurring for only short periods of time, they have the potential to ruin the pump if the NPSHa falls below the NPSHr and cavitation occurs.

# Total Dynamic Suction Lift

Some manufacturers show the total dynamic suction lift [TDSL] on their pump curves. This value is a measure of the suction lift and total pressure losses allowable on the suction side of the pump before the pump will cavitate, *assuming the pump is installed at sea level*. TDSL of 15 feet indicates that the total of static lift, all friction losses, and water vapor pressure can be no more than 15 feet, *if the pump is operating at sea level*. If the pump operates at a higher elevation, the TDSL is less than that shown on the pump curve.

TDSL can be confusing because of the implication that it is the vertical distance that the pump can lift the water and still work. However, this description is true only if all the friction losses in the suction line are zero. Obviously suction line losses are never zero, so the actual allowable static suction lift is always something less than the TDSL shown on the pump curve. The maximum static suction lift (for which the pump will still perform adequately) depends on the following:

- atmospheric pressure
- absolute pressure the pump requires at the inlet [NPSHr]
- friction losses in the pipe and fittings between the water surface and the impeller

The TDSL value is accurate only at sea level. Because the NPSHr value is the same at any elevation, it is the preferred value to use rather than the TDSL.

## Cavitation

Cavitation is the phenomenon resulting from the pressure somewhere in the pump being below the saturation vapor pressure for water. Saturated vapor pressure of water is the pressure of the vapor phase (steam) in the liquid. When the saturated vapor pressure equals the pressure of the liquid, conversion to steam begins. At atmospheric pressure, this occurs at 100°C or 212°F and is called boiling. However, it is a function of both temperature and pressure, so if the pressure is lowered far enough boiling can occur at any temperature above freezing. If the pressure in a pump falls below the saturated vapor pressure, vapor bubbles appear. Bad things can happen to pumps when water is allowed to turn into vapor and then back to liquid. The main problem is that water vapor occupies about 1,000 times as much space as water in the liquid form. If water is allowed to vaporize (form bubbles), when it recondenses to liquid it causes very high local velocities and water hammer on the surface, which can lead to severe damage. This phenomenon is known as cavitation. If cavitation occurs in a centrifugal pump, it usually occurs on the backside of the vane and can completely destroy the vane.

The results of cavitation can be dramatic. Cavitated impellers can be full of little holes as if made by a cutting torch. The rim of the impeller is often eroded to the point that the original impeller diameter no longer exists. In the longer term, the erosion of the impeller results in an unbalanced impeller, and the bearings will be destroyed. In extreme cases damage can occur in a very short time. If cavitation is suspected, the pump should be shut down immediately before it is permanently damaged.

Cavitation damage on a boat propeller is shown in figure 5-4, and cavitation damage on an impeller is shown in figure 5-5. Cavitation causes noise; it is often described as sounding like stones or gravel are passing through the pump. To prevent cavitation, designers need to take caution to ensure that the inlet pressure is not below NPSH required.

Figure 5-4 (left)
*Propeller cavitation damage (Reprinted from Wikipedia Commons.)*

Figure 5-5 (right)
*Impeller cavitation damage (Reprinted from Bell Metallurgy.)*

# Pressure Within the Pump

Figure 5-6 shows a plot of pressure typical in centrifugal pumps. The pressure at the inlet to the suction line is 8 feet of water. As the water passes through the suction line and fittings into the impeller there are losses due to friction and sudden bends. As it enters the vane area, it is accelerated, but energy hasn't been added yet. This causes the pressure to drop a little more so that the minimum is reached in the eye or on the first part of the vane. If the pump head falls below the vapor pressure line for the water temperature, cavitation can occur. Hence, at sea level, cavitation will not occur until the water temperature is slightly below 120°F. However, at 2,000 feet elevation cavitation can occur at both temperatures shown. It would appear that cavitation could occur at 2,000 feet elevation for water temperatures above about 80°F.

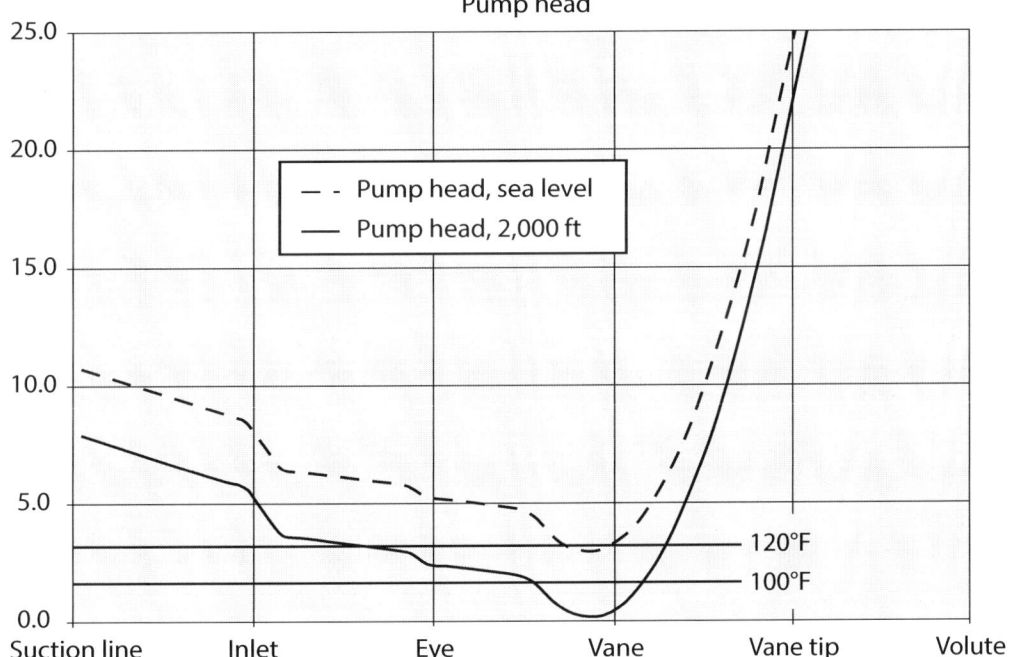

Figure 5-6
*Absolute pressure within a pump*

Chapter 5: Pump Performance

Some people erroneously think that cavitation can be caused by air entering the suction line. Although air entering the suction line reduces performance, it does not cause cavitation, and prevention of air leakage alone will not prevent cavitation. Simply stated, if NPSHa is insufficient (less than NPSHr), the pump can cavitate.

## Pump Curves

For a pump to adequately do the job required, it must be able to deliver the correct amount (flow rate) of water, and also it must produce the correct head (or pressure). The pump curve for that particular make and model of pump will help determine whether or not a particular pump will perform correctly.

Without a doubt, the most important part of system design is matching the pump (and its power unit) to the irrigation system that it will serve. Far too often, people try to match the system to the pump, and this almost always results in problems. Careful design and survey information is necessary so that the correct flow rate of water at the correct pressure can be delivered to the system. The friction losses through the system, plus the elevation differences, must be taken into account when calculating the TDH required at the pump. At this point a pump can be selected that will perform well at the desired conditions.

In many cases, the system requirements are not constant but can vary over a range of flow rates and pressures. Examples of this would be turf or golf course systems with various sizes of zones, wheel roll systems in which the number of sprinklers operating at any one time varies, or center pivot systems with corner arms.

When constant speed electric motors are used as a driving unit, the importance of careful pump selection cannot be overemphasized. With internal combustion engines, when the pump, power unit, and system are not exactly matched, the speed of the power unit may be changed slightly to compensate for mismatching; but the power unit must have sufficient brake horsepower output at the new operating speed. The speed of standard alternating current electric motors cannot be changed unless a VFD is incorporated into the system.

In many cases, the systems are complicated and a single pumping unit may not perform all the duties required. In these cases, booster pumps (pumps in series) or multiple pumps (pumps in parallel) may be required. For both booster pumping systems and parallel pumping systems, starting and stopping the pumping units can be automated. Careful attention must be paid to installing the necessary valves on any system where more than one pump is used. Pump stations using multiple pumps or VFD-controlled pumps are often used for golf courses and other complex pumping situations.

# Centrifugal Pump Curves

Normal horizontal centrifugal pump curves contain a great deal of information as shown in figures 5-7 and 5-8.

The most apparent set of curves is the Q-H or flow-head curves, which show the pressure (head) that a pump can produce at any particular flow rate. Generally, these curves slope down and to the right with the maximum head normally being at a flow rate of zero (shutoff head).

VFDs allow electric motors to operate at various speeds, but pumps driven by conventional electric motors operate at a fixed rotational speed {rpm}. Therefore, the pump curves for pumps connected to electric motors show the data for a constant speed (nominally 1,800 or 3,600 rpm), and different Q-H curves are included for different impeller diameters with the largest possible impeller diameter located at the top (see fig. 5-7).

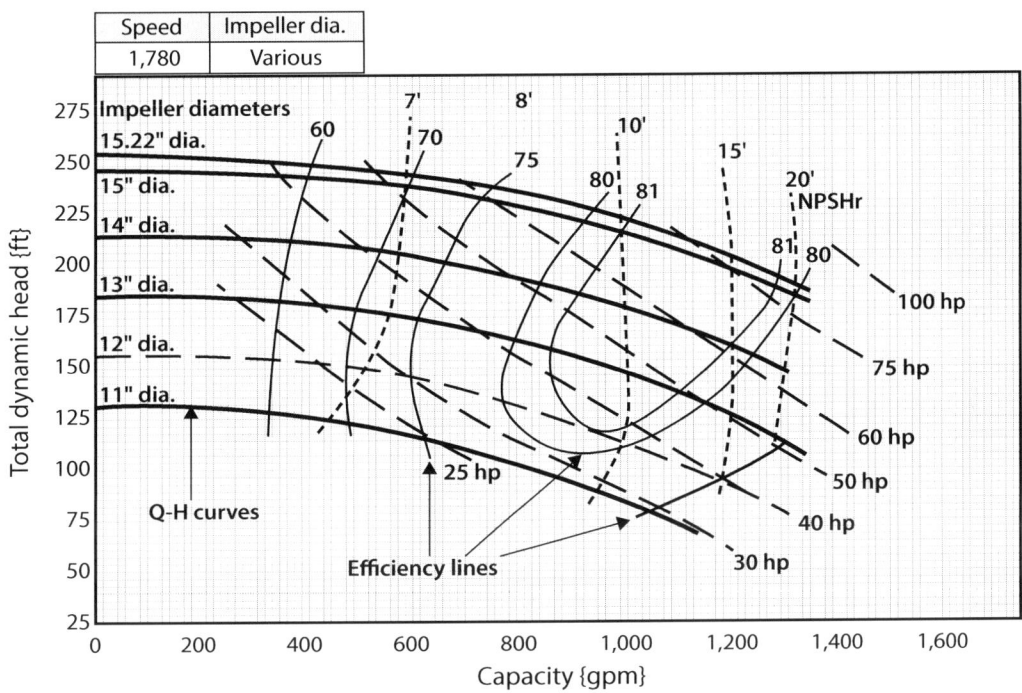

**Figure 5-7**
Pump curve for a fixed speed (electric motor) powered pump

Almost any impeller diameter between the maximum and minimum shown can be used for any particular pump. Impellers can be trimmed and balanced at a machine shop to have a diameter somewhere between the largest and smallest diameters shown on the curve. The pump affinity laws (discussed in chapter 6) can be used to predict the performance for diameters not shown on the curve. In some cases interpolation between published curves is reasonable. Impeller diameters should never be used that are less than the minimum or more than the maximum diameters shown on the published curves.

If the pump is to be connected to an internal combustion engine, which can operate at different speeds, the pump curve applies to a fixed impeller diameter (the full or maximum diameter as stated on the curve). Different Q-H curves are shown for different pump speeds, the maximum recommended speed being the top curve of the set (see fig. 5-8).

**Figure 5-8**
*Pump curve for a variable-speed (engine) powered pump*

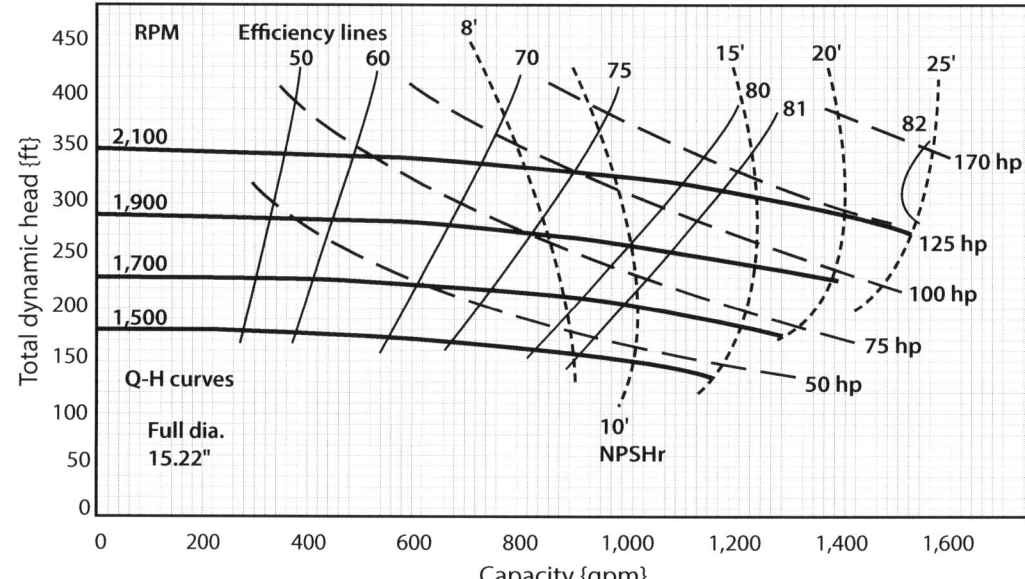

The pump should never be operated at a speed higher than that shown on the pump curve or at a speed higher than that recommended for the engine being used. For belt drive pumps, the maximum speed may be less than the maximum speed shown on the curve but, if so, that is noted on the pump curve.

The next basic piece of information that can be obtained from the pump curve is the pump efficiency, the percentage of the power delivered to the pump that is converted to useful power applied to the water. Most centrifugal pump curves show efficiency lines for various efficiencies (e.g., 75, 77, 80 percent). If the operating point is between two efficiency lines, the actual efficiency is interpolated.

A third set of lines on the pump curve are the brake horsepower lines. Once the operating point is known, the power required to operate this pump at that point can be estimated by interpolating between lines. Because the interpolation may be broad, between 75 horsepower and 100 horsepower for instance, the actual power required cannot be estimated accurately. It would be more accurate to calculate the brake horsepower using equation 4-3.

**Equation 4-3**
*Brake horsepower*

$$Bhp = \frac{Whp}{E_{pump}} = \frac{H \{ft\} \times Q \{gpm\}}{3{,}960 \times E_{pump}}$$

where

$Whp$ = water horsepower {hp}
$E_{pump}$ = pump efficiency converted to a decimal (e.g., 76% = 0.76)
$H$ = total dynamic head {ft}
$Q$ = flow rate {gpm}

The horsepower lines on the pump curve are useful as a preliminary estimate for electric motor sizing. The motor size can be selected by choosing the horsepower line *above* the operating point (not the horsepower line *closest* to the operating point). In most cases the horsepower line increments shown on pump curves match the common sizes of electric motors available (i.e., 30, 40, 50, 60, 75, 100, 125, 150, 200 horsepower, etc.)

Net positive suction head required is very important in pump selection and design. Most pump curves show the NPSHr either as approximately vertical lines similar to the efficiency lines or as a flow vs. NPSHr curve on the top or bottom of the pump curve. In either case, this indicates the minimum NPSHr that must be present to ensure that the pump does not cavitate. Generally, the NPSHr increases significantly as the flow rate increases.

## Vertical Turbine Pump Curves

Normal vertical turbine pump curves contain the same information as horizontal centrifugal pumps but usually display the information in a slightly different manner (see figs. 5-9 and 5-10). Also, they show the information *for one stage only* (e.g., one bowl assembly). The users must modify these values depending on the number of stages. Again, the Q-H curves generally slope down and to the right with the maximum head normally being at a flow rate of zero (shutoff head). However, vertical turbine Q-H curves sometimes flatten, or even turn upward, at the midpoint of the curve.

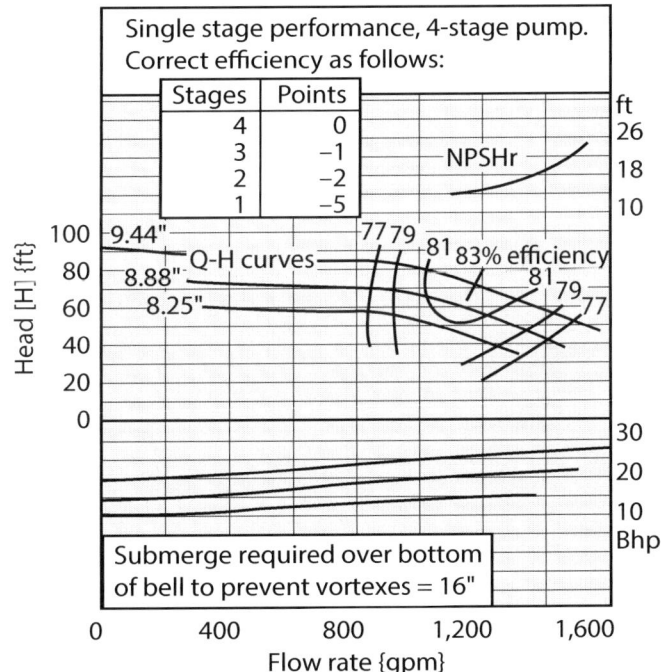

**Figure 5-9**
*Vertical turbine pump curve #1*

Because most vertical turbine pumps are connected to electric motors, turbine pump curves usually are drawn for a fixed pump speed (nominally 1,800 or 3,600 rpm). Different Q-H curves are shown for different impeller diameters, the largest possible impeller diameter being the top curve of the set. If the pump is connected to an internal combustion engine, a right-angle gear drive is used (see fig. 5-11). Whether the gear head has a 1:1 ratio or something else, the pump speed must be known so that the correct pump curve can be used. Pump affinity laws can be used to predict vertical turbine pump performance at speeds other than 1,800 or 3,600 rpm.

The pump efficiency information for a vertical turbine pump is shown in the same manner as for centrifugal pumps (see fig. 5-9) or as an efficiency vs. flow rate curve with one curve for each impeller diameter shown (see fig. 5-10). The efficiency curves may be above or below the Q-H curves and have their own scale on the vertical axis.

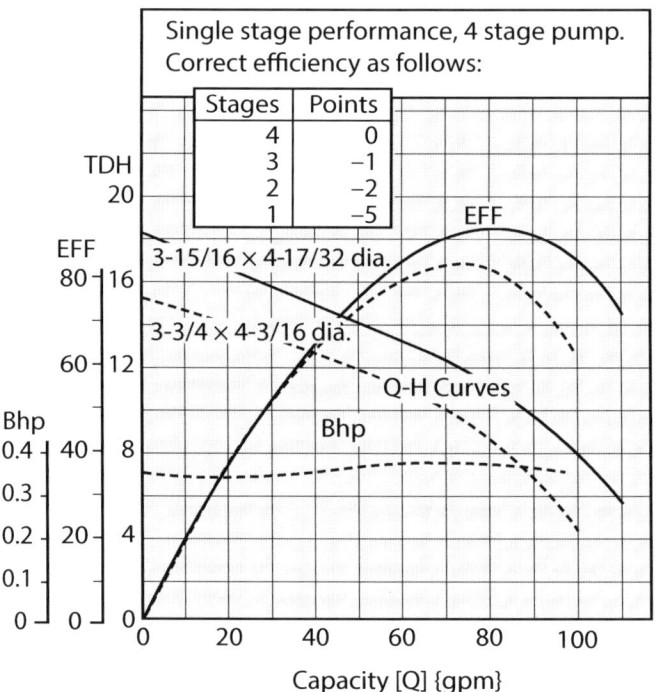

**Figure 5-10**
*Vertical pump curve #2*

**Figure 5-11**
*Internal combustion engine, right-angle gear drive, and vertical turbine pump*

Because most turbine pumps will be installed with a multiple number of impellers, the efficiency values shown usually assume four or more impellers. In normal operation, the first one or two impellers are less efficient because the water has to be channeled from the water source into the first bowl. The bowls then direct the water efficiently to the next stage. If the pump has only one or two stages, the overall efficiency will not be as high as the same pump with five or six stages. As a result, most vertical turbine pump curves indicate how much the efficiency (and sometimes the head) should be reduced if fewer than the assumed minimum numbers of stages are used.

A third set of lines shown on vertical turbine pump curves are the brake horsepower lines. The brake horsepower curves are normally shown as brake horsepower vs. flow curves, with one curve for each impeller diameter, and utilize their own vertical scale. The brake horsepower shown is the brake horsepower required to power one stage. The total brake horsepower required is the brake horsepower for one stage, multiplied by the number of stages. As with centrifugal pumps, calculating the power required with equation 4-3 is more accurate than using the pump curves. Using the efficiency and total head produced by all stages, equation 4-3 gives the total power required for all stages.

Because submergence can be an issue with vertical turbine pumps, the minimum submergence required is often shown on vertical turbine pump curves (see fig. 5-9) or elsewhere in the pump manufacturer's catalog. Minimum submergence is the distance the intake of the pump must be below the surface of the water so that vortexes do not form. If vortexes do form, the pump draws in air, which reduces the capacity of the pump and causes air entrapment problems in the pipelines of the irrigation system.

## Pump Selection

Before beginning the pump selection process, the various operating points that will be encountered need to be clearly defined. Then the designer identifies a pump, or a combination of pumps, that can meet the range of operating points expected.

Most pump selection is a process of examining catalogs and considering the possible choices. Although thorough catalog shopping is time-consuming, it is a relatively simple process once the various makes and models become familiar. Most pump manufacturers make a wide variety of pumps for many applications, only some of which are suited to irrigation applications. Usually only two or three manufacturers are considered. The choice of a pump may be simplified by the owner's desire to purchase certain unique features or to work with a certain dealer and/or manufacturer. Part of the purchase is reliability. Reliability and serviceability can limit the selection to a few common brands and/or dealers.

There is seldom a perfect choice for any application. Each choice has advantages and disadvantages. In the selection process it is advisable to keep notes on the choices and the pros and cons of each choice. Following are some of the questions that should be answered in the selection process:

- Will this pump deliver the entire range of required flows?
- Will this pump deliver the entire range of required pressures?
- What is the efficiency at each operation situation?
- Would the power unit size change if the efficiency were better?
- Is the NPSHr acceptable?
- Does the pump operating speed match the reasonable speed for the intended power unit?
- Are the inlet and outlet fittings compatible with the system?
- Are costs — initial, operating, and maintenance — acceptable? (However, selection should never be based on cost alone.)
- Would availability and serviceability be problems?
- Does this application have any unique aspects?

The individual selecting the pump should have a clear understanding of the priority of each of these items to the system owner and operator. In some situations, a pump capable of pumping a wide range of flows is more important than efficiency. If the flow range is relatively narrow and the hours of use per year are high, the primary selection criteria may be efficiency/energy costs.

Most of the major pump manufacturers have entire catalogs on computer disk or available via the Internet. These catalogs are usually coupled with a selector program that prints the choices for any given flow and pressure requirements. The software prints a pump curve with the design flow and head marked on the curve. This software quickly narrows the choices to just a few models. It is always desirable to print several choices and carefully compare them before making a final selection. The choice should always be confirmed by a careful examination of the published pump curve for the pump selected. The computer-generated pump curve alone should not be relied on.

Usually the pumping requirements, as specified by flow and head, do not appear as a point exactly on one of the published pump curves. It is impossible for manufacturers to publish curves for all situations. Most pump models encountered in irrigation applications have been manufactured for many years, and there is a large bank of test data confirming what is shown in the published curves. Pump tests done throughout North America have shown that the manufacturers' published curves usually represent pump performance very accurately. Still, they do not cover all situations. In cases where the pump curves do not cover a specific pump speed or impeller diameter, the use of pump affinity laws can be useful.

# Practice Questions

1. What would be the design flow rate for a system irrigating 12 acres with an irrigation efficiency of 75 percent, an $ET_O$ of 0.24 inches per day, landscape coefficient of 0.7, and no effective rainfall? If the system is only run 12 hours in a 2-day period, what would be the design flow rate?

   _____

2. What is the total dynamic head for a system flowing 200 gpm in 300 feet of Class 200 PVC (velocity = 4.98 ft/s, $H_f$ = 0.86 psi/100 feet), with 50 feet of elevation lift, discharge pressure requirement at the end of the pipe of 40 psi, and total suction head of 15 feet?

   _____ ft

3. What causes cavitation?

   _____

   _____

   _____

   _____

Chapter 5: Pump Performance

# Chapter 6

# Families of Curves

## Learning Objectives

The following objectives are the focus of chapter 6:
- understand how to use affinity laws
- understand pumps in series
- understand pumps in parallel

## Impeller Diameter and Speed Families of Curves

It is standard practice to publish a set of curves associated with a particular pump designation. Generally there are two families of curves shown. For a fixed speed pump (e.g., electric, close-coupled) there will be multiple impeller diameters shown. The nominal speed will be 1,800 or slightly less or 3,600 or slightly less. For example, a Berkeley centrifugal irrigation pump designated B4Z_H shows nominal 1,750 rpm and impeller diameters from 7.5 to 9.5 inches. However, a B4Z_H variable speed pump curve shows a diameter of 9.5 inches and speeds from 1,200 to 2,800 rpm. The respective families of curves are shown in figures 6-1 and 6-2.

**Figure 6-1**
*Family of pump curves at nominal speed*

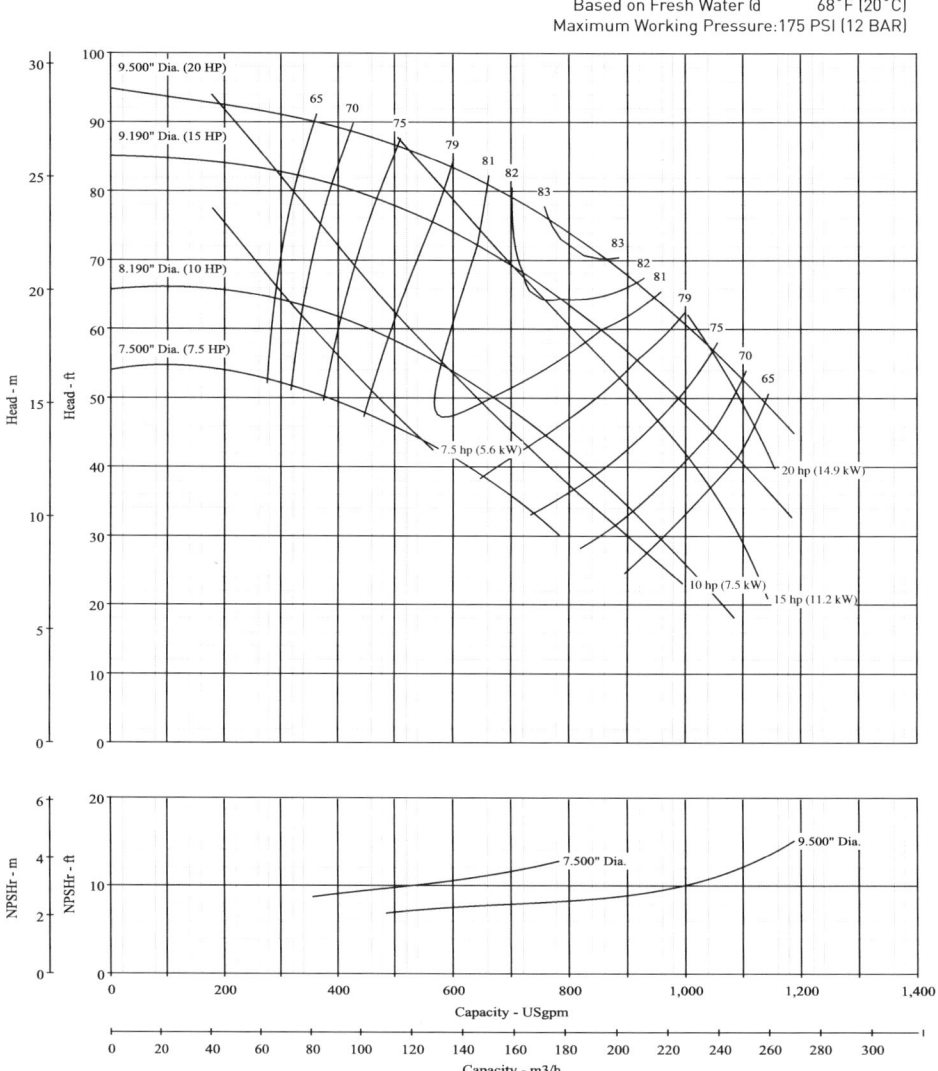

This set of curves shows how performance varies with impeller diameter. Most pump curves contain at least six essential pieces of information for any given capacity. First, there is a curve representing the capacities (flows) and head available for a given impeller diameter. The upper curve in figure 6-1 shows that the impeller diameter is 9.5 inches. Selecting a capacity of 800 gpm shows that the B4Z_H will produce about 73 feet of head. The efficiency is about 83 percent. The horsepower (brake horsepower, Bhp) is slightly less than 20 horsepower (about 18 brake horsepower). NPHSr is about 9 feet. Using equation 4-3 to do a quick check of horsepower equals the following: Bhp = (73 × 800) ÷ (3,960 × 0.83) = 18 Bhp.

The variable speed family of curves shows similar information but specific to speed rather than to impeller diameter. Note that NPSHr is shown for maximum speed.

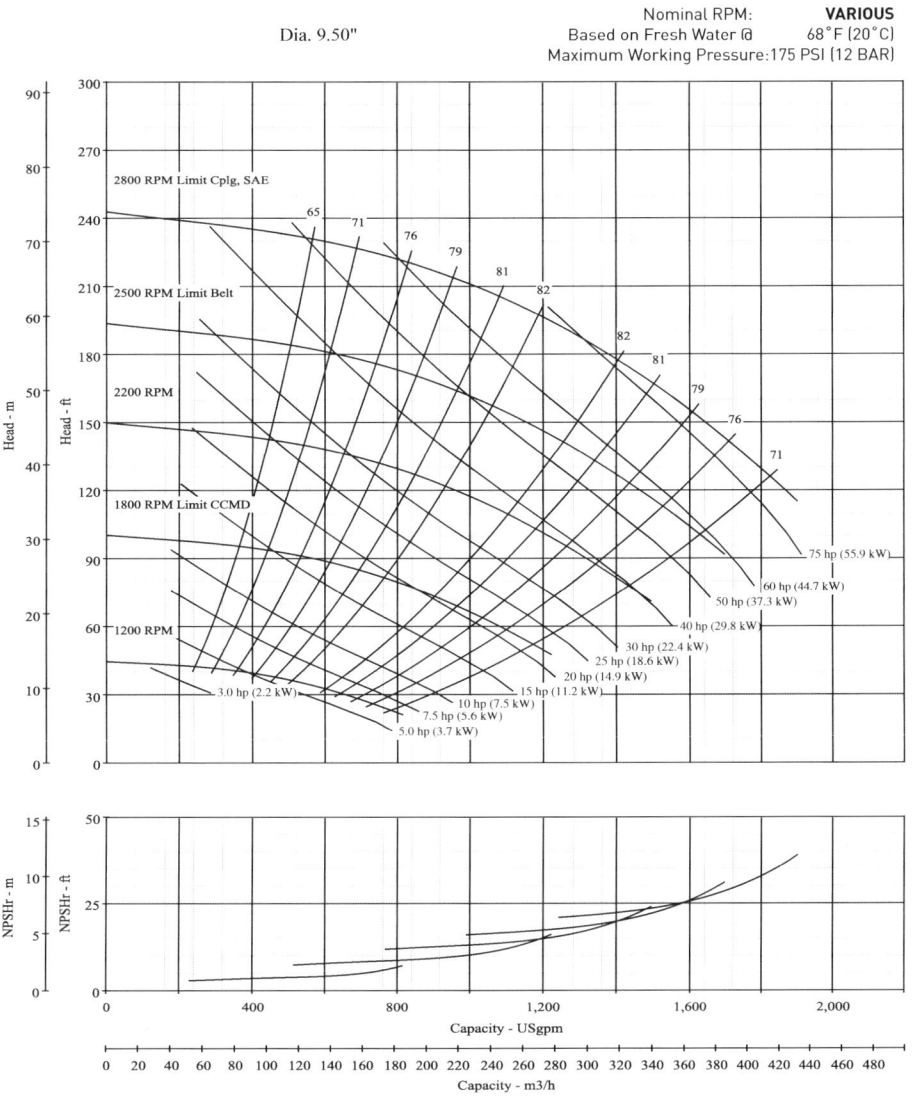

**Figure 6-2**
*Family of pump curves at given impeller diameter*

## Affinity Laws

At times it may be useful to estimate the performance of a pump at a speed other than those shown on the chart or for impeller diameters other than those shown. There is a set of equations known as the affinity laws that allow calculation of head, pressure, and power for speeds or diameters not known. They all involve the ratio of the parameter changed (speed or diameter) to a power, and the powers are 1, 2, and 3. It may be helpful to recall some relationships in order to understand the affinity laws. First remember that the tip velocity of the impeller is the product of the speed and impeller radius (half of the diameter). Flow is area multiplied by velocity, or *flow goes with velocity*. Second, remember that the energy equation is in terms of head. The kinetic energy component, velocity, is squared in the equation, so *head goes as velocity squared*. Third, remember that power is the product of head and velocity. Therefore, flow is proportional to the change in vane tip velocity whether it be a result of rotational speed change or impeller diameter, head is proportional to velocity squared, and power is proportional to the product of the two. *Power is proportional to velocity cubed*. The affinity laws are shown below.

# Changing the Pump Speed

In some situations, a pump curve is not published for the required pump speed. The most common example is when the published curve is for a 1,750-rpm electric motor and the power unit is an internal combustion engine designed to operate at a different speed. To estimate the performance of the pump at something other than 1,750 rpm, pump affinity laws are used.

For change in rotational speed [N], use equation 6-1.

**Equation 6-1**
*Change in rotational speed*

$$\frac{Q_1}{Q_2} = \frac{N_1}{N_2} \qquad \frac{H_1}{H_2} = \frac{(N_1)^2}{(N_2)^2} \qquad \frac{P_1}{P_2} = \frac{(N_1)^3}{(N_2)^3}$$

This relationship shows that
- flow rate varies directly as the speed.
- head (pressure) varies as the square of the speed.
- power required varies as the cube of the speed.

These affinity laws are based on the basic assumption that the pump efficiency does not change as the speed is varied. This assumption holds if the speed changes are relatively small (less than 10 percent).

## Changing the Impeller Diameter

Changing the speed of the driving shaft is not always possible, such as when an electric motor is used or when it is necessary to run an internal combustion engine at a certain speed. An example is when the engine is also used to run a three-phase generator for an electric drive center pivot. The engine speed must remain constant to ensure the generator produces the correct frequency of alternating current. When the pump speed of rotation cannot be changed, changing the impeller diameter can alter the pump performance. This is normally done by physically trimming the impeller to a smaller diameter. Impeller diameters are variable within certain limits as shown on the pump curve. Obviously the casing has a maximum-sized impeller that fits in it. This size is usually shown on the original pump curve as the full impeller. At a certain reduced diameter, there is too much backflow around the sides of the impeller back into the inlet. This backflow results in poor efficiencies for impellers that are small relative to the volute. Impellers should not be trimmed to diameters smaller than the smallest shown on the published pump curve (see fig. 6-1).

If the diameter changes are relatively small (less than 10 percent), pump affinity laws show that the flow rate [Q], TDH [H], and power [P] are affected by pump impeller diameter [D] according to the following relationships.

For change in impeller diameter [D], use equation 6-2.

$$\frac{Q_1}{Q_2} = \frac{D_1}{D_2} \qquad \frac{H_1}{H_2} = \frac{(D_1)^2}{(D_2)^2} \qquad \frac{P_1}{P_2} = \frac{(D_1)^3}{(D_2)^3}$$

**Equation 6-2**
*Change in impeller diameter*

This relationship shows that
- flow rate varies directly as the diameter.
- head (pressure) varies as the square of the diameter.
- power required varies as the cube of the diameter.

**Example 6-1**
*Operating conditions*

Assume that the operating conditions demonstrated for the B4Z_H pump are changed to use an 8.75-inch diameter impeller. What are the operating conditions?

**Solution**

$$Q_2 = Q_1 \times \frac{D_2}{D_1} = 800 \times \left(\frac{8.75}{9.5}\right) = 737 \text{ gpm}$$

$$H_2 = H_1 \times \left(\frac{D_2}{D_1}\right)^2 = 73 \times \left(\frac{8.75}{9.5}\right)^2 = 62 \text{ ft}$$

$$P_2 = P_1 \times \left(\frac{D_2}{D_1}\right)^3 = 18 \times \left(\frac{8.75}{9.5}\right)^3 = 14 \text{ hp}$$

Check horsepower. Looking at the curve, efficiency is estimated to be 81 percent.

$$\text{Bhp} = \frac{H \times Q}{3{,}960 \times E} = \frac{62 \times 737}{3{,}960 \times 0.81} = 14 \text{ hp}$$

A typical application of the affinity laws is in the case of a power unit that needs to run faster than the pump to give the desired performance.

Another example of when impeller trimming may be useful is when an irrigator previously obtained water from an open water source (canal or dugout) that was later replaced by a closed pipeline under pressure. Because there is now more pressure at the pump, less head needs to be added and the impeller can be trimmed. The affinity laws determine how much to trim the impeller. This option is usually more cost-effective than buying a smaller pumping unit.

Many pump manufacturers now supply software that helps with affinity law calculations. Some of these programs also print a curve for the proposed speed and impeller diameter. These computer-generated curves are a result of calculations and are subject to the cautions mentioned previously in this section.

## Multiple Pump Stations

Often it is not possible for a single pump to deliver the desired performance over all expected conditions. If that is the case, multiple pumps are a possible option, arranged in parallel (see fig. 6-3a), series (see fig. 6-3b), or in rare cases a combination of both.

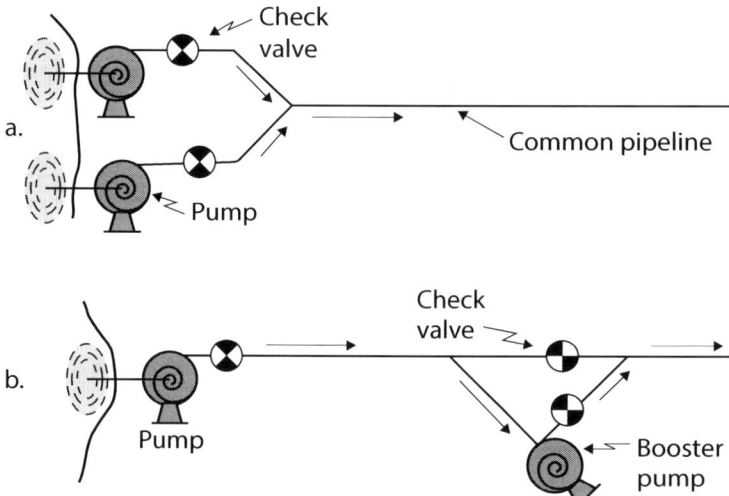

**Figure 6-3**
a. Parallel pump setup
b. Series pump setup

With pumps in parallel, the flow produced by each pump is additive, but both pumps must produce the same pressure. With pumps in series, the pressure produced by each pump is additive, but the flow through both pumps is the same. In some cases a booster pump can be used to add pressure to only a portion of the total flow (e.g., when a part of the irrigation system is at an elevation significantly higher than the rest of the system). Knowing the system curve and the individual pump curves for the pumps connecting in either series or parallel, the total pump curve for the combination of pumps can be generated and the following can be predicted:

- flow and pressure when any one pump is running
- flow and pressure when any combination of pumps is running

## Pumps in Series

If the required pressure is too high for one pump, two or more pumps may be placed in series to act as booster pumps and increase the total head produced. This is analogous to the successive stages in vertical turbine pumps except that now each pump has its own power unit. In this case the flow through each pump is the same, and the total head produced by the pumps in combination is equal to the sum of the heads produced by each individual pump (see fig. 6-4). Each pump does not have to produce the same head, but each pump must be capable of pumping the same flow rate unless the booster pump pumps only a portion of the total flow.

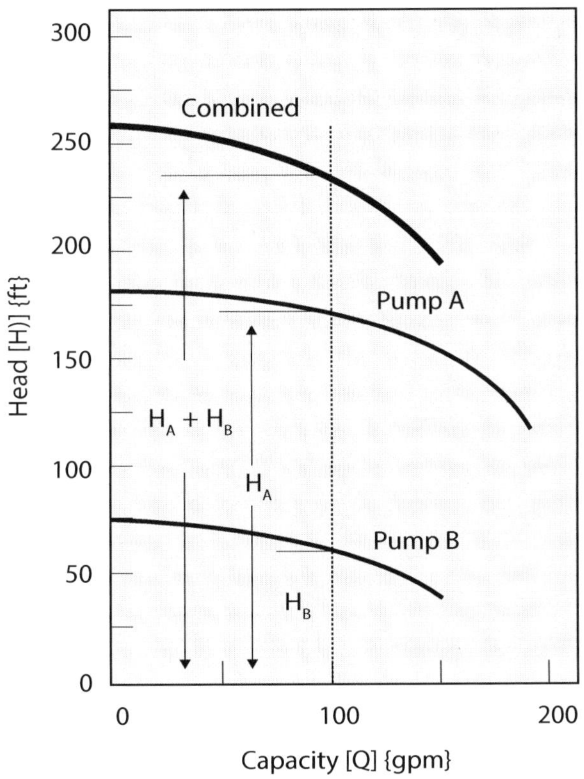

**Figure 6-4**
*Curves for pumps in series*

Pumps in series do not need to be installed at the same location. One example is a pump site where one pump lifts water from a river to a field above the river at which point a second (booster) pump further pressurizes all or some of the water to operate the sprinklers correctly. If both pumps (or one large pump) were located at the river, the pressure in the pipeline at the start of the system would be greater, and higher pressure rated pipe may have to be used. Another example is where only part of the irrigated area is at a high elevation. One pump supplies enough pressure to irrigate most of the area while a booster pump, installed at some point in the system, adds extra pressure to only the water that is flowing to the higher area. In this case the flow rate of the second pump would be much less than the first pump, and the energy/fuel consumed would be for only the energy to boost that smaller flow rate of water rather than the total flow.

## Pumps in Parallel

If the required flow rate is too high for one pump, two (or more) pumps may be connected in parallel to produce the desired flow rate. In this case each pump is producing the same head/pressure. The total flow is the sum of the flows from each pump (see fig. 6-5). The pumps need not pump the same quantity of water, but each pump must produce the same head. Often, different sized pumps are connected in parallel when the flow rates vary widely (e.g., large turf, golf course systems). When the flow is small, a small pump operates. When the flow increases, a larger pump starts and the smaller one shuts down. When the maximum flow is required, both pumps operate together.

**Chapter 6: Families of Curves**

**Figure 6-5**
*Curves for pumps in parallel*

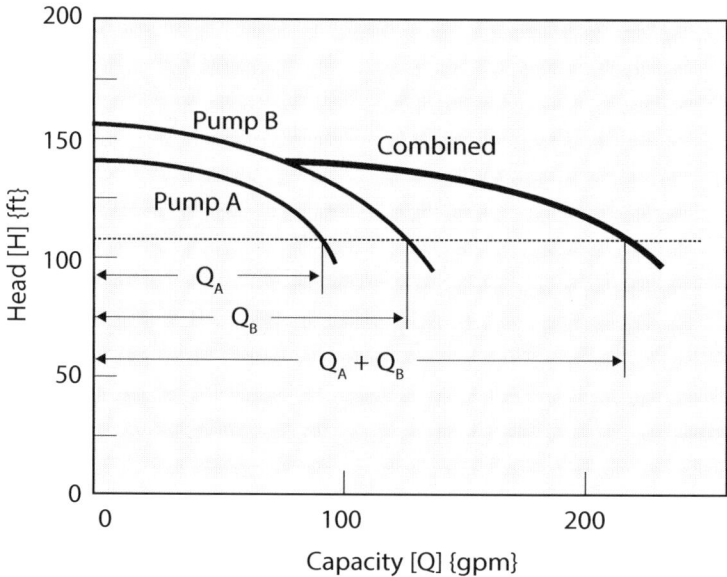

For the parallel pump example shown in figure 6-5, the shutoff head for pump B is more than the shutoff head for pump A, which presents an interesting situation at low flows. Note that the "combined" Q-H curve does not start until the head is less than the shutoff head for pump A. When the head is more than the shutoff head for pump A, pump A produces no flow (assuming it has a check valve installed in the discharge line). If it does not have a check valve, then pump B would actually pump water backward through pump A. A pump should not be operated against its shutoff head because no water flows, and the pump can be damaged from overheating. Figure 6-6 shows an installation with several pumps in parallel.

**Figure 6-6**
*Multiple pumps in parallel in an irrigation district*

# Practice Questions

1. According to the affinity laws, flow changes with what power of the speed?

2. According to the affinity laws, head changes with what power of the speed?

3. According to the affinity laws, power changes with what power of the speed?

4. With pumps in series, which is the same and which is additive (head and/or flow)?

5. With pumps in parallel, which is the same and which is additive (head and/or flow)?

# Chapter 7

# Pump Selection

## Learning Objectives

The following objectives are the focus of chapter 7:
- know the pump selection procedure
- know how to use total dynamic head and design flow rate to pick a pump
- calculate net positive suction head
- introduce computer pump selection

## Matching Selection Criteria

The task in choosing a pump is to find a pump that provides the desired capacity and total head while operating at close to maximum efficiency — that is, while using the least possible horsepower and energy and incurring the lowest possible demand charges.

## Total Dynamic Head

Total dynamic head consists of everything affecting pressure. The pumping lift is the distance in feet required to raise the water to the ground surface. The pump discharge pressure is the pressure needed by the sprinklers or emitters in addition to the pressure needed to overcome elevation differences and pressure losses caused by friction. Losses must be included and are discussed in detail later.

## Selection Procedure

The following are steps taken when choosing a pump:
- Estimate the needed capacity and total dynamic head.
- Use a catalog of manufacturer's pump performance curves to find a pump that will provide the needed capacity and total head at close to maximum efficiency.
- Use the pump performance curve to determine the horsepower requirement of the pump.
- Select the appropriate electric motor or engine. The engine revolutions per minute should be close to maximum engine efficiency.

# System Requirements

Proper centrifugal pump selection is a fairly simple process that begins with knowledge of the following system requirement data:

- flow rate
- total dynamic head

## Flow Rate

When sizing pumps and reading pump curves, two pieces of information are needed: flow rate and total dynamic head. The flow requirement is generally expressed in gallons per minute. Flow is the sum of all outlets that will be operating at the same time while the pump is operating. If the system is divided into blocks, zones, or stations that operate separately, then the flow consists of the maximum flow of any one station.

**Example 7-1**
*Flow rate example*

A soccer field has six rows of full-circle sprinklers. Each row has four sprinkler heads, and each head uses 20 gpm. Two rows of sprinklers operate at a time. What is the flow requirement of the pump?

**Solution**

- 4 (heads) × 2 (rows) = 8 sprinkler heads operating at once
- 8 (heads) × 20 (gpm) = 160 gpm

Therefore, the pump will need to deliver 160 gpm.

If flow changes, so does head (pressure). The relationship between the two will be shown by the pump characteristic curves. The actual point of operation is determined by pump and system characteristics.

## Total Dynamic Head

Care must be taken to ensure that the components of TDH are in the same units. Normally sprinkler or emitter pressure is expressed in pounds per square inch, but lift is in feet. Most Q-H curves are flow {gpm} vs. head {ft}. As discussed in chapter 2, one foot of head of water is equivalent to 0.433 psi, and one pound per square inch of pressure is equivalent to 2.31 feet of head of water.

TDH is one of the most important factors in the pump selection process. It is a function of the system. TDH is pressure that is generally expressed in feet, and its value is the summation of the following items:

- suction pipe friction loss
- suction lift
- suction entrance loss
- discharge pipe friction loss
- discharge lift
- sprinkler operating pressure
- miscellaneous fittings loss

The following outlines the procedure for gathering and computing system TDH using the example shown in figure 7-1.

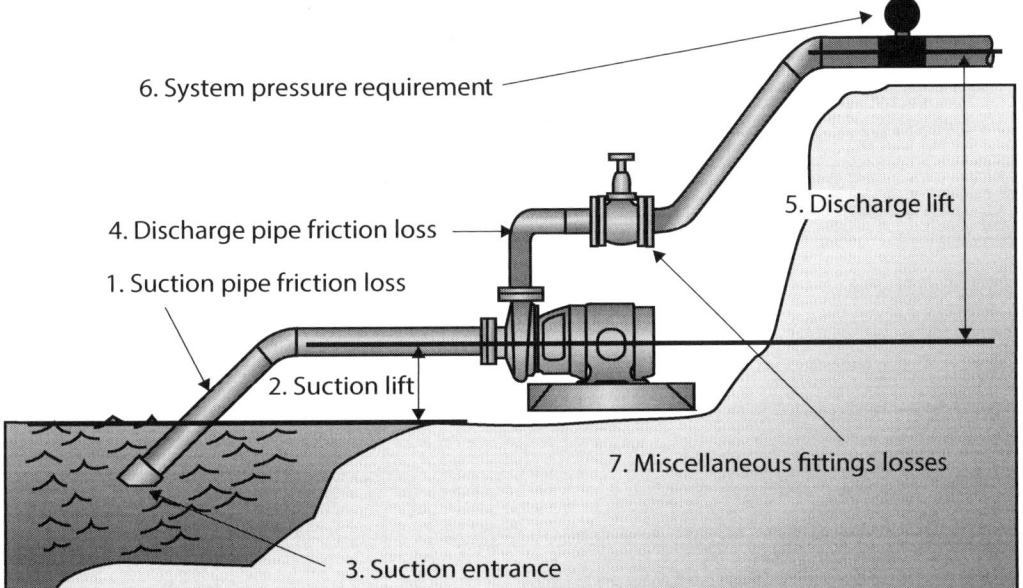

Figure 7-1
*Determining TDH*

Find total dynamic head for a capacity of 320 gpm.

Example 7-2
*TDH example*

### Solution

| | | |
|---|---|---|
| 1. Suction friction (6-inch steel pipe, 20 feet long) | = | 1 ft |
| 2. Suction lift | = | 5 ft |
| 3. Suction entrance loss | = | 2 ft |
| 4. Discharge friction (6-inch steel pipe, 1,000 feet long) | = | 14 ft |
| 5. Discharge lift | = | 15 ft |
| 6. Pressure required for system operation | = | 100 ft |
| 7. Miscellaneous fitting losses | = | 5 ft |
| **TDH** | = | 142 ft |

The following examines each of the items involved with TDH as shown in figure 7-1.

1. *Suction pipe friction loss* — This is the friction loss that occurs as water moves through the entire length of suction pipe. It's determined by the flow rate and the size and type of pipe installed.

   The example system has 20 feet of 6-inch diameter steel pipe. From friction tables, this is 0.28 feet, which is rounded up to 1 foot.

2. *Suction lift* — This is simply the vertical distance in feet from the surface of the water to the centerline of the pump. The example has a suction lift of 5 feet.

3. *Suction entrance loss* — To determine suction entrance loss, figure the foot valve loss from tables depending on size, then add an extra foot for velocity head entrance loss.

The example has a 1-foot loss for the foot valve, plus an extra foot for velocity head. This totals 2 feet. Velocity head is calculated with equation 7-1.

**Equation 7-1**
*Velocity head*

$$H_v = \frac{v^2}{2 \times g}$$

where
- $g$ = acceleration due to gravity $\{32.2 \text{ ft/s}^2\}$
- $v$ = velocity $\{\text{ft/s}\}$
- $H_v$ = velocity head $\{\text{ft}\}$

4. *Discharge pipe friction loss* — To determine discharge pipe friction loss, use the same procedure as for suction pipe friction loss. The only difference is that it is on the other side of the pump.

   The example system has 1,000 feet of 6-inch diameter steel pipe. Using the same friction tables to calculate this, the result is 14 feet of head loss.

5. *Discharge lift* — Discharge lift is similar to suction lift, except on the discharge side. It's the vertical distance in feet from the centerline of the pump to the highest point in the system.

   The example has a discharge lift of 15 feet.

6. *System pressure requirement* — To determine the system pressure requirement, find the pressure needed at the end of the line and convert to feet of head. In the example, the amount needed is 43 psi, which converts to 100 feet.

   The calculation is as follows:

   psi × 2.31 = ft

   43 × 2.31 = 100 ft

7. *Miscellaneous fittings losses* — This accounts for losses in any valves, elbows, filters, or other fittings a system may have. This information can also be obtained from friction tables.

   The example has 5 feet of miscellaneous losses. Adding all these together equals the TDH requirement of 142 feet.

It can't be overstressed that TDH is the most important factor in the pump selection process. It's the pressure the pump will work against to keep it on its performance curve while moving liquid through the system.

The goal is to select a pump that will meet the system's TDH requirement on its performance curve and where it produces the desired flow rate. These conditions should be met while simultaneously optimizing the efficiency of the pump.

# Pumpage

Pumpage is the water being pumped. The following further describes different pumpage properties that can affect the pump and should be considered during the selection process.

## Chemical Makeup

Is the liquid corrosive or noncorrosive? This affects pump materials of construction and the shaft sealing method. For example, some injected particles are corrosive.

## Temperature

Is the liquid hot or cold? This also affects pump materials of construction and the shaft sealing method. It also affects the net positive suction head available.

## Specific Gravity or Density of Liquid

Is the liquid heavy or light? This affects pump performance and horsepower requirements. For the most part, in irrigation the pumping water will have a density of 1.0, or a specific weight of 62.4 pounds per cubic feet.

## Presence of Abrasives or Solids

Does the liquid contain sand or silt? Are there any solids or stringy materials as commonly found in sewage, liquid manure, or process waste? This affects pump model, impeller type, operating speed, and materials of construction.

## Viscosity

Is the water thick like molasses or clean and clear? This affects pump performance and horsepower requirements.

## Presence of Gases

Does the liquid have entrained gases? This greatly affects pump performance.

## Vapor Pressure

The vapor pressure of water is the pressure it takes to keep the water from evaporating at a given temperature. At boiling temperature (212°F) the vapor pressure is one atmosphere or about 33.9 feet of water. At normal irrigation water temperatures it is less than 2 feet. Figure 7-2 shows vapor pressure of water with temperature.

**Figure 7-2**
*Vapor pressure of water with temperature*

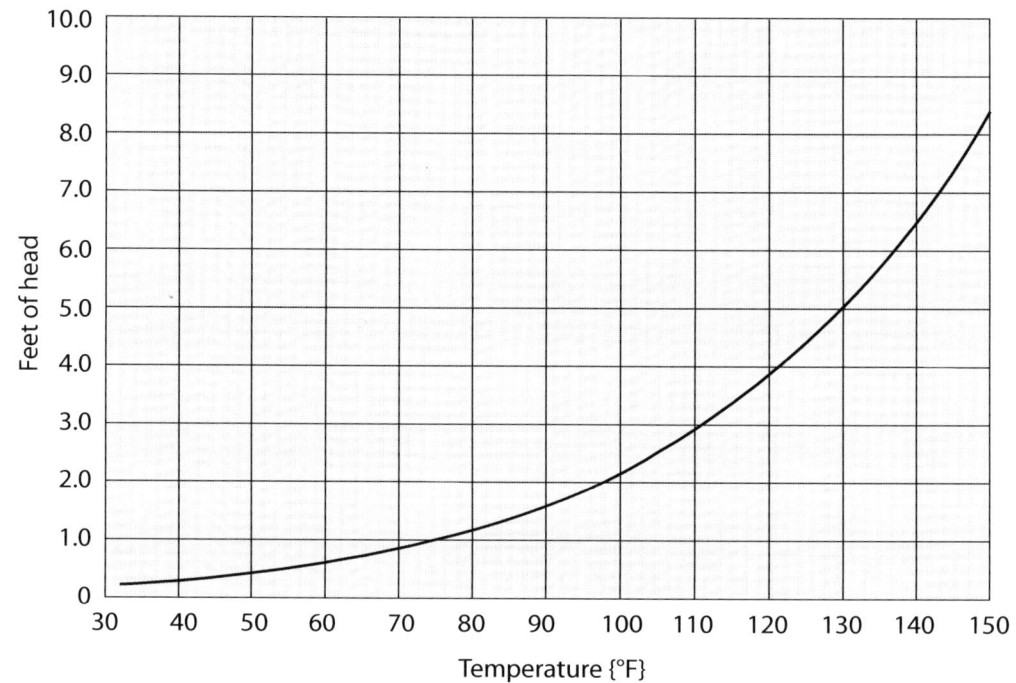

In the example, the system will be pumping clear water at 60°F. The properties of water at this temperature will not be a major concern in pump selection.

# Net Positive Suction Head

Most pump problems have to do with the suction side of the pump. Usually the cause is not taking into account the net positive suction head required [NPSHr] by the pump. NPSH available [NPSHa] is a function of the earth's atmospheric pressure. It is generally expressed in feet, and with water at sea level, 33.9 feet or 14.7 psi is available. It is important to be certain that the atmospheric pressure will exceed the suction losses and "push" the water into the pump.

NPSH also plays a major role in proper pump selection. NPSHa from the system must exceed the NPSHr by the pump, preferably by at least 2 feet. This is subject to changes in

- altitude or barometric pressure.
- static height of liquid above or below pump centerline.
- all suction-side friction and entrance losses.
- liquid's vapor pressure at pumping temperature.

NPSH calculations involve both atmospheric pressure as determined by altitude of the installation and temperature of the water. The altitude determines the total pressure available $[H_a]$ and is given by table 7-1.

Table 7-1
Atmospheric pressure $[H_a]$

| Altitude {ft} | Water {ft} | psi |
|---|---|---|
| Sea level | 33.9 | 14.7 |
| 1,000 | 32.8 | 14.2 |
| 2,000 | 31.6 | 13.7 |
| 3,000 | 30.5 | 13.2 |
| 4,000 | 29.4 | 12.7 |
| 5,000 | 28.3 | 12.2 |
| 6,000 | 27.3 | 11.8 |
| 7,000 | 26.2 | 11.3 |
| 8,000 | 25.2 | 10.9 |

The temperature of the water determines the vapor pressure $[H_{vp}]$ and it is given by table 7-2.

Table 7-2
Vapor pressure of water $[H_{vp}]$

| Temperature | | Vapor pressure {ft} |
|---|---|---|
| °F | °C | |
| 50 | 10.0 | 0.41 |
| 60 | 15.6 | 0.59 |
| 70 | 21.1 | 0.84 |
| 80 | 26.7 | 1.17 |
| 90 | 32.2 | 1.61 |
| 100 | 37.8 | 2.19 |
| 110 | 43.3 | 2.95 |
| 120 | 48.9 | 3.91 |
| 130 | 54.4 | 5.14 |
| 140 | 60.0 | 6.68 |

The two tables can be combined to give an expression equivalent to atmospheric pressure $[H_a]$ less vapor pressure $[H_{vp}]$, known as net pressure, thereby simplifying the calculations. Table 7-3 shows net pressure $(H_a - H_{vp})$.

Chapter 7: Pump Selection

**Table 7-3**
*NPSHa {ft} for given altitude and temperature*

| Alt {ft}\T {°F} | 40 | 50 | 60 | 70 | 80 | 90 | 100 | 110 | 120 | 130 |
|---|---|---|---|---|---|---|---|---|---|---|
| 0 | 33.6 | 33.5 | 33.3 | 33.1 | 32.7 | 32.3 | 31.7 | 31.0 | 30.0 | 28.8 |
| 500 | 33.0 | 32.9 | 32.7 | 32.5 | 32.1 | 31.7 | 31.1 | 30.4 | 29.4 | 28.2 |
| 1,000 | 32.4 | 32.3 | 32.1 | 31.9 | 31.5 | 31.1 | 30.5 | 29.8 | 28.8 | 27.6 |
| 1,500 | 31.8 | 31.7 | 31.5 | 31.3 | 30.9 | 30.5 | 29.9 | 29.2 | 28.2 | 27.0 |
| 2,000 | 31.2 | 31.1 | 30.9 | 30.7 | 30.4 | 29.9 | 29.3 | 28.6 | 27.6 | 26.4 |
| 2,500 | 30.7 | 30.5 | 30.4 | 30.1 | 29.8 | 29.3 | 28.8 | 28.0 | 27.0 | 25.8 |
| 3,000 | 30.1 | 30.0 | 29.8 | 29.5 | 29.2 | 28.8 | 28.2 | 27.4 | 26.5 | 25.3 |
| 3,500 | 29.5 | 29.4 | 29.2 | 29.0 | 28.7 | 28.2 | 27.6 | 26.9 | 25.9 | 24.7 |
| 4,000 | 29.0 | 28.9 | 28.7 | 28.4 | 28.1 | 27.7 | 27.1 | 26.3 | 25.4 | 24.2 |
| 4,500 | 28.5 | 28.3 | 28.1 | 27.9 | 27.6 | 27.1 | 26.6 | 25.8 | 24.8 | 23.6 |
| 5,000 | 27.9 | 27.8 | 27.6 | 27.4 | 27.0 | 26.6 | 26.0 | 25.3 | 24.3 | 23.1 |
| 5,500 | 27.4 | 27.3 | 27.1 | 26.8 | 26.5 | 26.1 | 25.5 | 24.7 | 23.8 | 22.6 |
| 6,000 | 26.9 | 26.8 | 26.6 | 26.3 | 26.0 | 25.6 | 25.0 | 24.2 | 23.3 | 22.0 |
| 6,500 | 26.4 | 26.2 | 26.1 | 25.8 | 25.5 | 25.0 | 24.5 | 23.7 | 22.8 | 21.5 |
| 7,000 | 25.9 | 25.7 | 25.6 | 25.3 | 25.0 | 24.5 | 24.0 | 23.2 | 22.3 | 21.0 |
| 7,500 | 25.4 | 25.3 | 25.1 | 24.8 | 24.5 | 24.1 | 23.5 | 22.7 | 21.8 | 20.5 |
| 8,000 | 24.9 | 24.8 | 24.6 | 24.3 | 24.0 | 23.6 | 23.0 | 22.2 | 21.3 | 20.1 |

Example 7-3 uses the data obtained in the TDH calculation to calculate NPSHa in the system shown in figure 7-1.

Net positive suction head available is calculated using equations 7-2a and 7-2b.

**Equations 7-2a and 7-2b**
*Net positive suction head available*

$$NPSHa = H_a - H_s - H_f - H_{vp}$$

$$NPSHa = (H_a - H_{vp}) - H_s - H_f$$

where

$NPSHa$ = net positive suction head available {ft}
$H_a$ = atmospheric pressure at elevation of water surface {ft}
$H_s$ = static lift between water surface and center line of pump {ft}
$H_f$ = suction line friction losses {ft}
$H_{vp}$ = vapor pressure of liquid being pumped {ft}
$H_a - H_{vp}$ = net pressure

Using data obtained in example 7-2, determine the NPSHa.

**Example 7-3**
*NPSHa calculation*

## Solution

|   |   |   | Eq. 7-2a | Eq. 7-2b | Source |
|---|---|---|---|---|---|
| 1. | Atmospheric pressure at sea level | $H_a$ | 34 | – | Table 7-1 |
| 2. | Net atmospheric pressure at sea level {60°F} | $H_a - H_{vp}$ | – | 33 | Table 7-3 |
| 3. | Static height of liquid below pump centerline | $H_s$ | 5 | 5 | Given |
| 4. | Suction pipe friction loss | $H_f$ | 1 | 1 | Calculated |
| 5. | Entrance loss | | 2 | 2 | Estimated |
| 6. | Suction side miscellaneous loss | | 1 | 1 | Estimated |
| 7. | Vapor pressure of water at {60°F} | $H_{vp}$ | 1 | – | Table 7-2 |
|   | **Total NPSHa** | | 24 | 24 | |

The solution using either equation 7-2a or 7-2b will give a correct answer, but most people will find it simpler and less prone to error to use table 7-3 and equation 7-2b.

Be conservative and adjust downward. The next section discusses how NPSHa is considered in pump sizing and selection.

# Reading Pump Curves

Pump performance curves provide a graphical representation of a centrifugal pump's performance characteristics based on water with a specific gravity of 1.0. A typical pump curve will indicate TDH, brake horsepower, efficiency, and NPSHr — all plotted in relation to the capacity range of the pump. The example calculations that follow use system requirement data of 47 gpm at 155 TDH.

The 1WC curve is a typical performance curve for a centrifugal pump (see fig. 7-3). Pressure values or TDH are indicated by reference point 1. TDH is usually shown in feet. Flow values (or capacity) are indicated by reference point 2. Capacity is usually shown in gallons per minute.

**Figure 7-3**
*Cornell 1WC pump curve (reference points 1–3)*

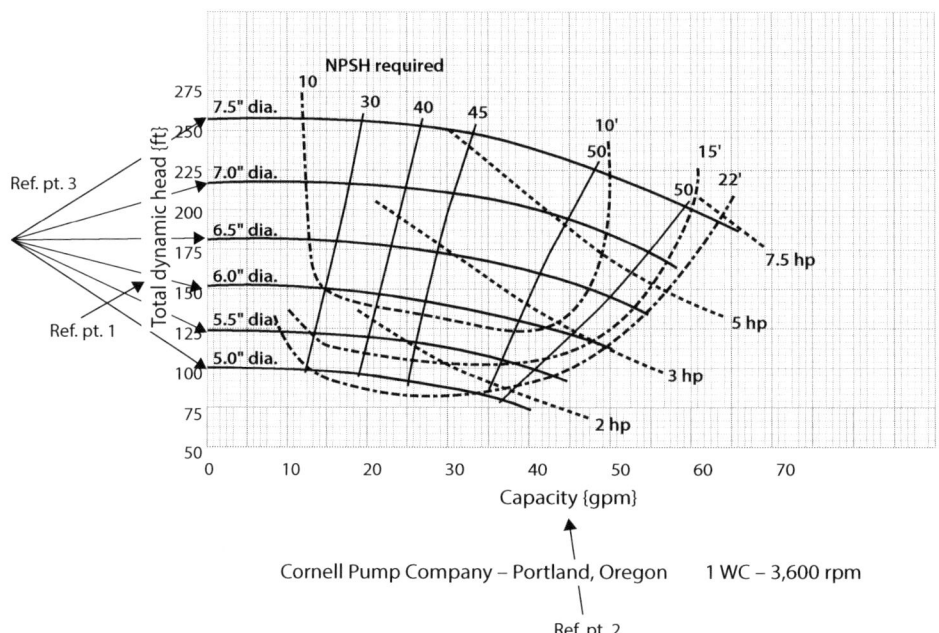

## Head Capacity Curve

The curve lines indicated by reference point 3 are the pump performance curves. They are plotted to indicate the pump's capacity in relation to total dynamic head. Each line reflects performance with different impeller trim diameters.

For this particular pump, 7.5 inches is the maximum impeller diameter available, and 5 inches is the minimum impeller diameter available. Impellers can generally be trimmed or machined in 1/16-inch increments to fine tune performance necessary to meet a system's requirements.

Notice that as the flow or capacity of a centrifugal pump increases, the TDH produced decreases.

## NPSHr Curve

The NPSHr curves are indicated by the vertical broken lines shown by reference point 4 (see fig. 7-4). The NPSH requirement value is determined by the pump manufacturer's hydraulic design. This value is usually indicated in feet. Notice that as flow or capacity of a centrifugal pump increases, NPSHr also increases.

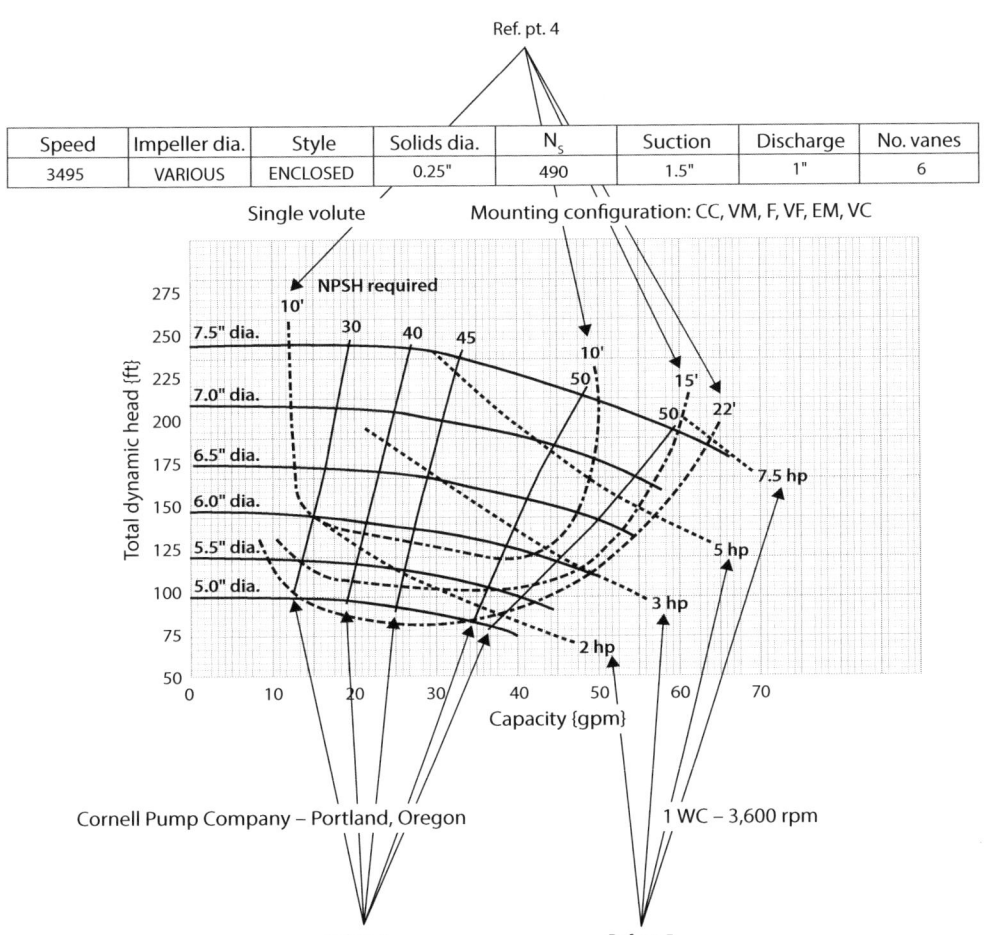

**Figure 7-4**
Cornell 1WC pump curve
(reference points 4–6)

## Horsepower Curve

The curves indicated by reference point 5 are horsepower curves. Each curve has a corresponding value indicating horsepower requirements of the pump at any given point relative to head capacity performance. The point where the head capacity curve and the horsepower curve intersect is where the pump will draw or require the respective horsepower value.

For example, the 6.5-inch impeller curve intersects the 3 horsepower curve at the approximate point where 30 gpm intersects 175 feet of head. Pump operation to the right increases the horsepower requirement; to the left decreases the horsepower requirement. The more flow produced by the pump, the more horsepower required by the pump.

The horsepower is calculated using equation 4-3. Noting that the efficiency is about 45 percent (reference point 6) for the 30 gpm and 175 feet of head point, the brake horsepower is $(175 \times 30) \div (3{,}960 \times 0.45) = 2.95$ or approximately 3 hp.

## Efficiency Curve

The curves indicated by reference point 6 in figure 7-4 are the pump's efficiency curves. Each curve has a corresponding value indicating the efficiency of the pump at any given point relative to head capacity performance. For this particular pump, efficiency ranges in value from 30 to 50 percent.

The pump's best efficiency point [BEP] is 50 percent, as indicated by reference point 7 (see fig. 7-5). BEP depends on hydraulic design by the pump manufacturer. It varies with pump model.

For example, the 6.5-inch diameter curve has a BEP of 50 percent at the intersection of 47 gpm at 155 TDH. The 7-inch diameter impeller head capacity performance curve line has a BEP of 32 percent at the intersection of 20 gpm at 220 TDH. Notice that as flow increases or decreases in relation to BEP, efficiency decreases.

Other information that can typically be found on a centrifugal pump performance curve includes impeller style, impeller diameter, specific speed, maximum diameter of solids, pump speed, pump model, and pump manufacturer (see figs. 7-5 and 7-6). How a manufacturer presents this information, if the information is presented, depends on the pump manufacturer. Following are brief explanations of each point:

- *Impeller style* (ref. point 8) — Different types of impellers are available depending on the application a pump is designed to handle. There are three basic types: enclosed, semi-open, and open. This particular pump has an enclosed impeller.
- *Impeller diameter* (ref. point 9) — The performance of this particular pump is plotted with a constant speed of 3,495 rpm, with various impeller trim diameters indicated by the various head capacity performance curves. Hence, the indication of the impeller diameter as "various."
  Some pump performance curves will indicate only a single impeller trim diameter with the head capacity performance curve lines indicating various pump speeds. Some pump performance curves will indicate both various impeller trim diameters and various pump speeds. This can be confusing. The key is to analyze each curve on its own to understand what the manufacturer is representing.
- *Specific speed* (ref. point 10) — Specific speed is indicated by the abbreviation Ns. This is easily confused with pump speed. It's a value that centrifugal pump design engineers use to classify impeller designs. Actually, the value doesn't mean much to the average person selecting a centrifugal pump for an irrigation application, but since most pump manufacturers indicate Ns on their curves, it does warrant some explanation.
- *Maximum diameter solids* (ref. point 11) — This indicates the maximum diameter spherical solid the pump impeller can pass without plugging.
- *Pump speed* (ref. point 12) — This indicates the shaft rotational speed in revolutions per minute on which the curve performance is based. If the performance curve is a variable speed curve, the information will indicate the variables similar to that shown by figure 7-5.
- *Pump model* (ref. point 13) — Every pump manufacturer has model designations. This particular pump model happens to be a Cornell model 1WC.

All pump curves for a given pump should indicate the pump manufacturer. Most centrifugal pump manufacturers are very helpful in answering questions concerning data presented on their curves. Any questions should be directed to the manufacturer or their local representative.

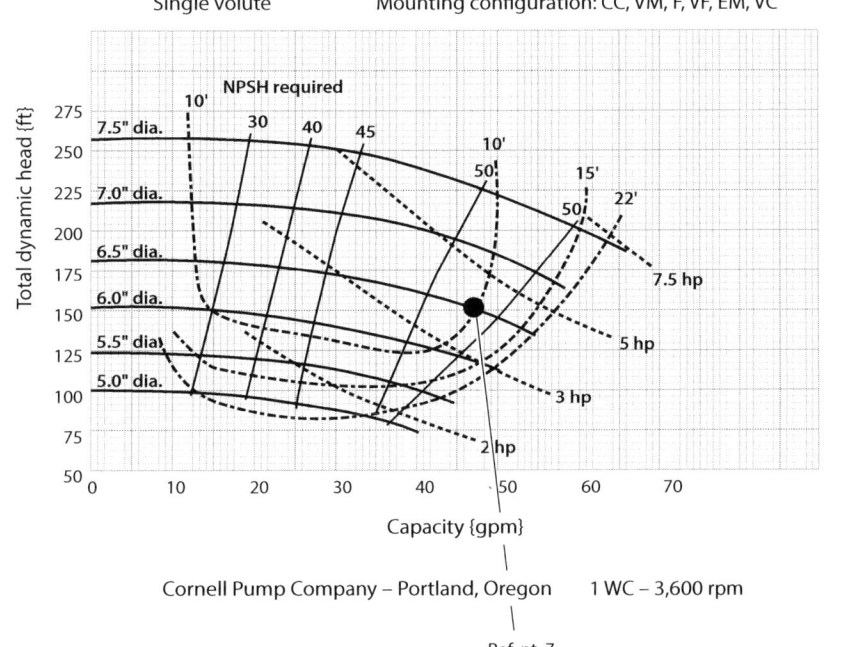

**Figure 7-5**
*Cornell 1WC pump curve (reference points 7–9)*

**Figure 7-6**
*Cornell 1WC pump curve (reference points 10-13)*

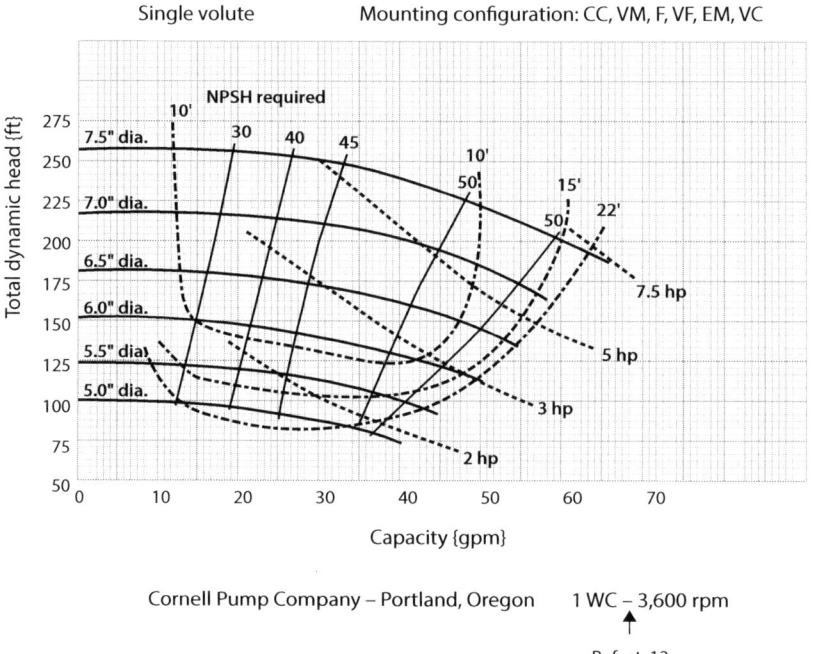

## Computer Programs for Pump Selection

There are several applications available to assist in pump selection. Perhaps the most widely used is Pump-Flo™ developed by Engineered Software, Inc. It is available at eng-software.com/products/pump-flo/. This software allows the user to select the manufacturer and then select a pump meeting the TDH, design flow, and NPHSa criteria. For example, the Pump-Flo™ program for Cornell Pumps is shown in figures 7-7 through 7-9 with progressive screen shots. The following selections have been made in figure 7-7:

- design flow rate of 550 gpm
- total dynamic head of 150 feet
- choose the operating point left of best efficiency point
- NPSHa = 15 feet
- clear liquid
- 1,800 rpm

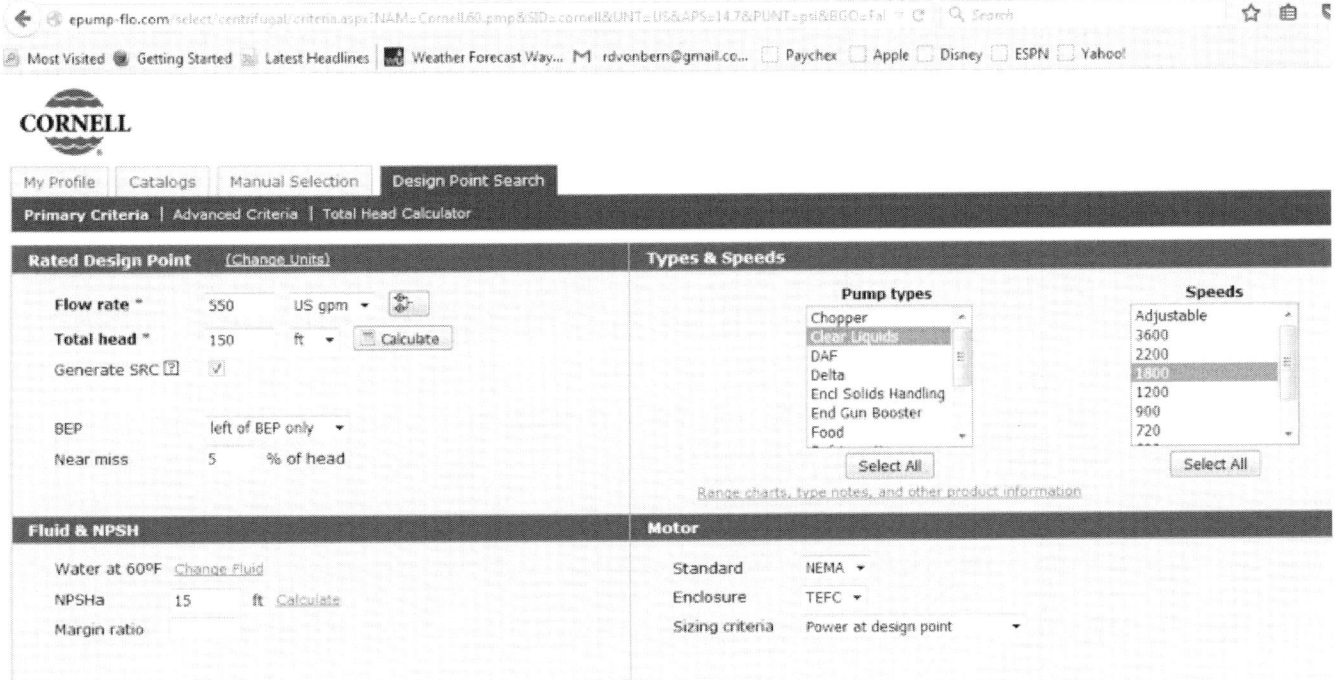

**Figure 7-7**
*Pump-Flo™ for Cornell Pumps for 550 gpm and 150 feet of head*

After clicking on the "Search" button, the screen shown in figure 7-8 comes up. It shows a possibility of nine different pumps. By clicking on the top one (4RB) the screen as shown in figure 7-9 comes up. This shows the 4RB pump with the appropriate trim (11.9375 inches) at 1,775 rpm (actual rpm versus nominal rpm). The operating point is shown by the chevron. The system curve (Pump-Flo™ calls it a resistance curve) is also shown, but users should be cautioned that it is developed primarily from pipe friction and does not include a fixed value for lift. Nonetheless, it shows an efficiency of about 75 percent, about 28 brake horsepower for 550 gpm, 150 feet of head, and about 8 feet of NPSHr. The motor horsepower is shown as 50 because that is the maximum horsepower on the design curve. If power at the design point had been selected, it would show 30 motor horsepower. If the user desires, a family of curves at different speeds can be developed by inserting the speeds in the "Multiple Speeds" (rpm) box and clicking "redraw." Speeds from 1,500 to 1,900 are shown in figure 7-10. These curves are based on the affinity laws.

Design Point: 550 US gpm, 150 ft.

| Preview | Type | Size | Curve | Speed (rpm) | Dia | Head (ft) | Eff (%) | BEP (%) | NPSHr (ft) | Power (hp) | Motor (hp) | Frame | Min flow (US gpm) | Impeller |
|---|---|---|---|---|---|---|---|---|---|---|---|---|---|---|
|  | Clear Liquids | 4RB | 4RB18 | 1775 | 11.9375 in | 150 | 75.4 | 85.5 | 8.03 | 27.6 | 50 | 326T | 150 | --- |
|  | Clear Liquids | 4HH | 4HH18 | 1780 | 12.125 in | 151 | 72.6 | 81 | 7.12 | 28.8 | 50 | 326T | 200 | --- |
| 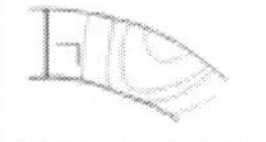 | Clear Liquids | 5RB | 5RB18 | 1780 | 11.875 in | 150 | 67.7 | 85.5 | 10 | 31.4 | 60 | 364T | 200 | --- |
| 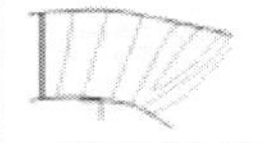 | Clear Liquids | 4HC | 4HC18 | 1780 | 12 in | 155 | 67.3 | 74.5 | 9 | 31.6 | 50 | 326T | 100 | --- |
|  | Clear Liquids | 5H | 5H18 | 1780 | 12.0625 in | 151 | 65.2 | 76.8 | 12 | 31.9 | 50 | 326T | 200 | --- |
|  | Clear Liquids | 6H | 6H18 | 1780 | 12.3125 in | 150 | 55.1 | 83.6 | 10 | 38.7 | 75 | 365T | 300 | --- |
| 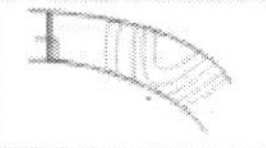 | Clear Liquids | 6RB | 6RB18 | 1780 | 12.3125 in | 151 | 51.6 | 89.1 | 12 | 59.8 | 100 | 405T | 408 | --- |
|  | Clear Liquids | 8H | 8H18 | 1785 | 12.25 in | 150 | 29.5 | 85.5 | 12 | 69 | 125 | 444T | 400 | --- |
| 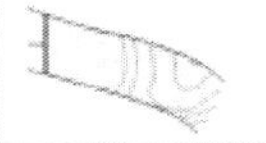 | Clear Liquids | 10RB | 10RB18 | 1785 | 12.81" × 13° | 150 | 24.2 | 85.4 | 11.7 | 110 | 150 | 445T | 500 | --- |

**Figure 7-8**
*Pump-Flo™ selection possibilities for 550 gpm and 150 feet of head*

A close review of figure 7-8 shows that the pump possibilities are arranged in decreasing efficiencies. However, none of the pumps operate under these conditions at the best efficiency point [BEP].

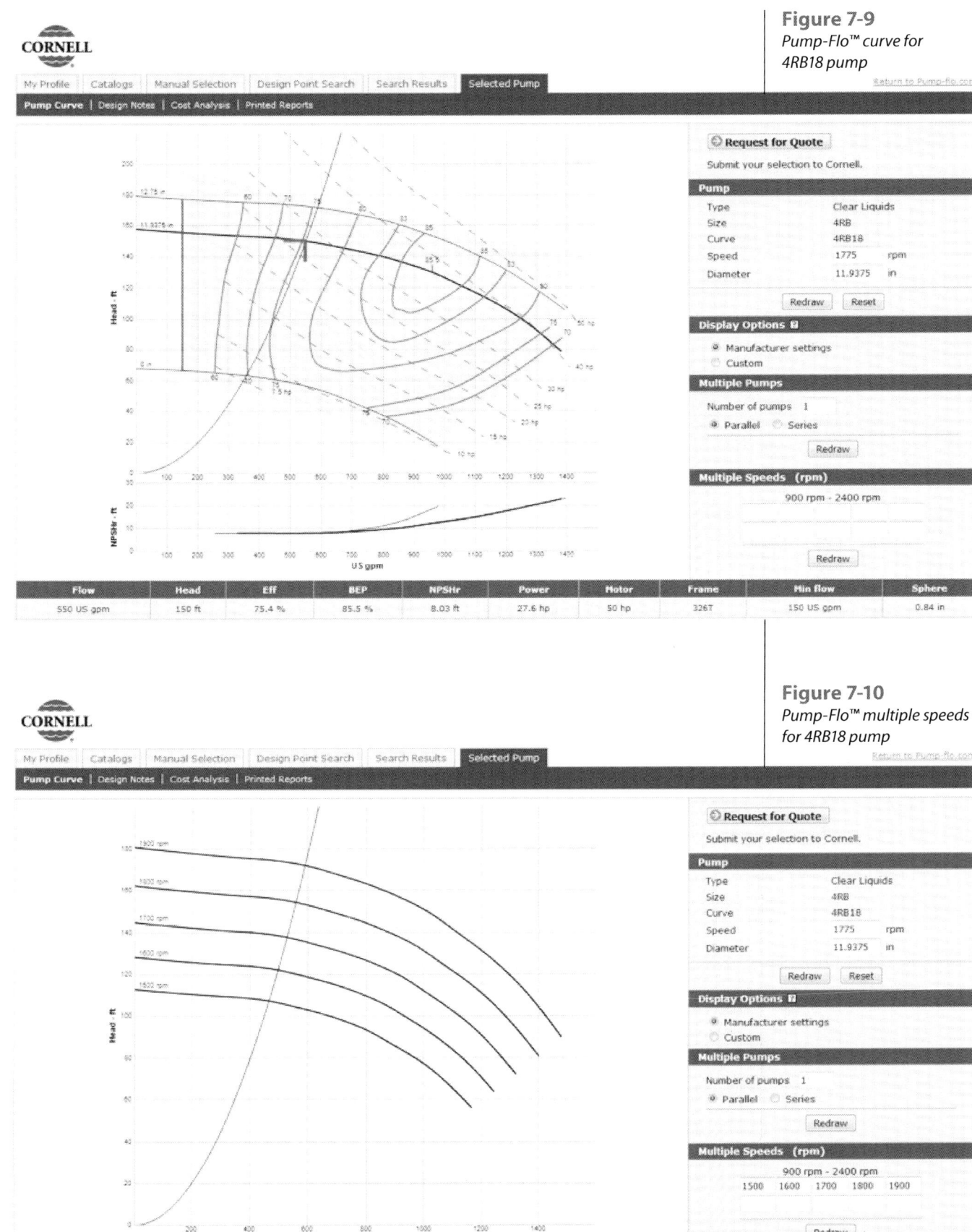

Figure 7-9
Pump-Flo™ curve for 4RB18 pump

Figure 7-10
Pump-Flo™ multiple speeds for 4RB18 pump

Chapter 7: Pump Selection

A similar search for a Grundfos pump using the same criteria produced figure 7-11, and similarly in figure 7-12 for Goulds pumps.

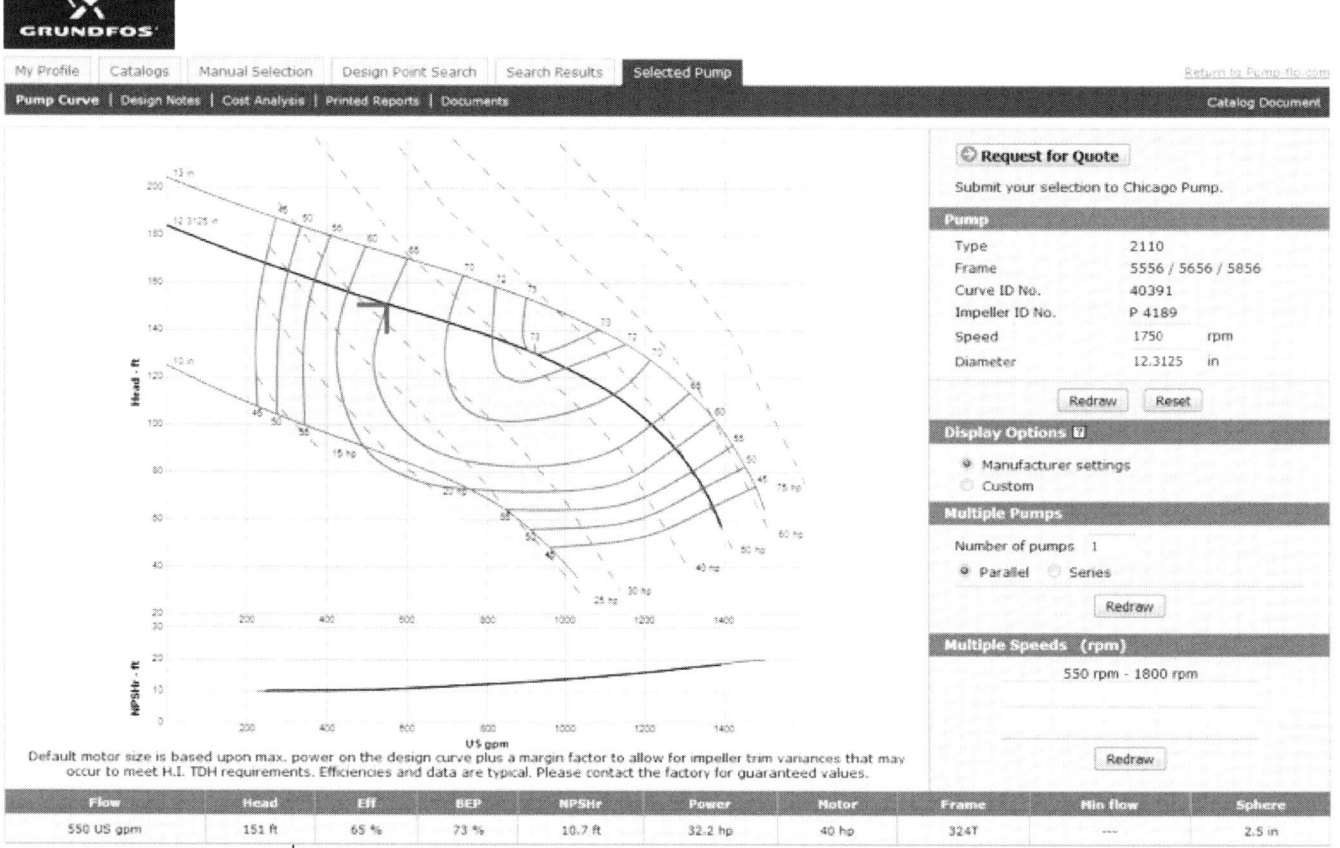

**Figure 7-11**
*Pump-Flo™ Grundfos pumps selection for 550 gpm and 150 feet of head*

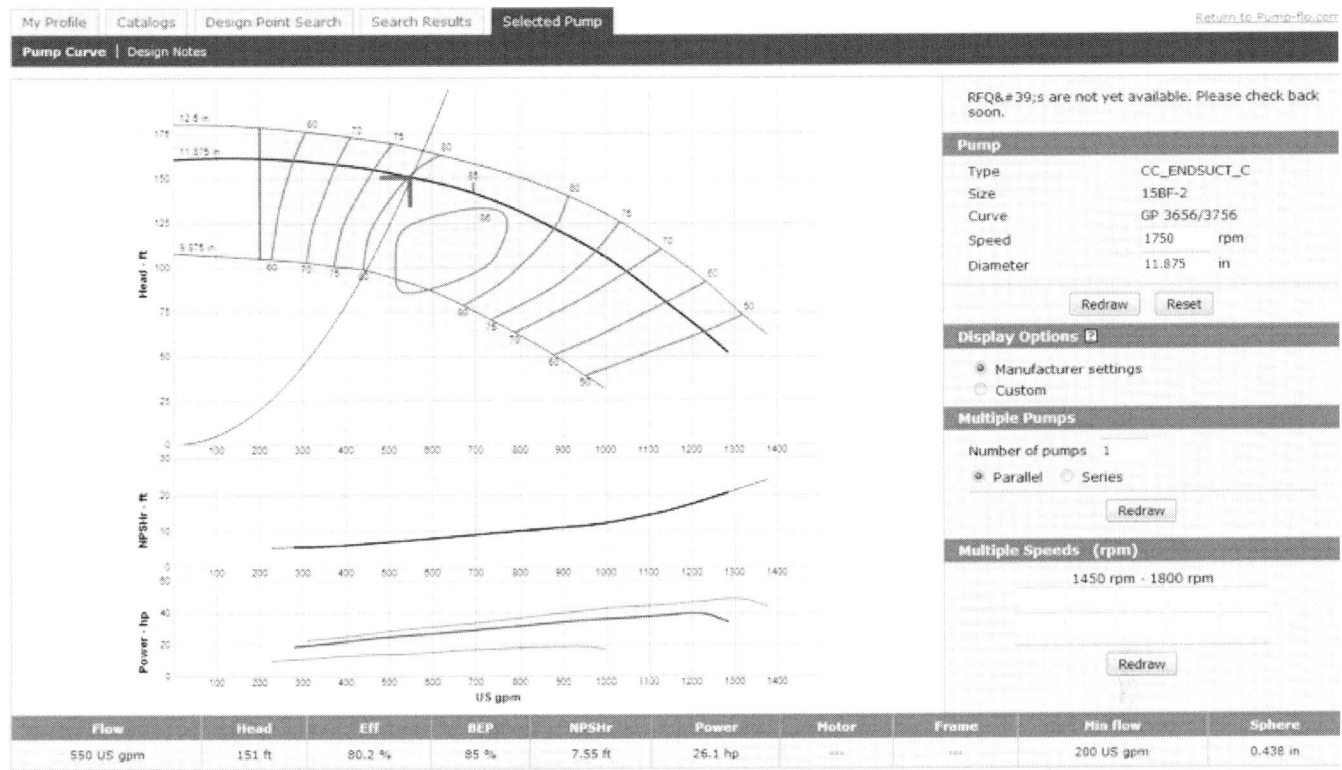

**Figure 7-12**
Pump-Flo™ Goulds pumps selection for 550 gpm and 150 feet of head

## Smart Phone Apps for Pump Selection

Technology changes rapidly, and it is difficult to stay up-to-date on available applications for smart phones. Users are advised to search for apps and evaluate them. One such app is the Mobile Tookit from Cornell Pumps. It includes calculators for TDH, NPSH, and friction loss, along with a product catalog. Rain for Rent and several others also offer free pump calculator apps.

Chapter 7: Pump Selection

# Practice Questions

1. What two values must be known to pick a pump?

    (1.) _____

    (2.) _____

2. Name five components of TDH.

    (1.) _____

    (2.) _____

    (3.) _____

    (4.) _____

    (5.) _____

3. How does altitude affect NPSHa?

    _____

    _____

    _____

    _____

4. How does water temperature affect NPSHa?

    _____

    _____

    _____

    _____

# Practice Questions *cont.*

5. Why is the suction side of a centrifugal pump generally larger than the discharge?

   _____

   _____

   _____

   _____

6. Name five items normally found on pump curves.

   (1.) _____

   (2.) _____

   (3.) _____

   (4.) _____

   (5.) _____

7. Using figure 5-7, find the head, efficiency, horsepower, and NPSHr for the pump running 1,000 gpm with a 15-inch impeller.

   SCRATCH PAD

   H = _____

   Efficiency = _____

   Bhp = _____

   NPSHr = _____

# Operating Point for Pumps

## Learning Objectives

The following objectives are the focus of chapter 8:
- understand the system curve
- determine pump operating point from system curve and pump curve

## Pump and System Curve Together

The actual performance of a pump is in reaction to the delivery and distribution system to which it is attached. Just as the pump can operate under many different flow and head conditions, so can the system. Any change in the distribution system (e.g., zone changes) will change the character of the system and will affect the pump. However, for a given system and pump, there is only one operating point. Exactly how the pump system and delivery system interact is the topic of this chapter.

## System Curve

A system curve involves several aspects of irrigation hydraulics. It is made up of everything that affects the total dynamic head under different flow conditions. As noted in chapter 7, total dynamic head includes suction pipe friction loss, suction lift, suction entrance losses, discharge pipe friction loss, discharge lift, sprinkler operating pressure, and miscellaneous fittings losses. Some of the factors of total dynamic head vary with flow, and some do not.

### Factors that Vary with Flow

*Suction Pipe Friction Loss and Suction Entrance Losses*

Both suction pipe friction loss and suction entrance losses vary with flow. However, the amount of loss, while significant in NPSHa calculations, does not significantly affect system curves.

## Pressure Head

Pressure head can change significantly as the number of discharge devices is changed, as would be expected with changes from one zone to another. Also, unless the discharge device is completely pressure compensated, there will be change in head with flow changes. This is likely the most significant change of all the components of the system curve. If the distribution system is a set of nonpressure regulated sprinklers, the relationship between head (pressure) and flow is given as equation 8-1.

**Equation 8-1**
*Relationship between head and flow*

$$Q = k \times C \times A \times \sqrt{p}$$

where

- $Q$ = sprinkler discharge {gpm}
- $A$ = cross-sectional area of nozzle {in.$^2$}
- $C$ = nozzle discharge coefficient depending on the size and construction of nozzle
- $p$ = pressure at the nozzle entrance {psi}
- $k$ = numerical constant

The effect of pressure on sprinkler discharge can be derived from equation 8-1. The discharge changes with the square root of the pressure change as shown in equation 8-2.

**Equation 8-2**
*Sprinkler discharge*

$$\frac{Q}{Q_0} = \sqrt{\frac{p}{p_0}}$$

where

- $Q$ = new sprinkler discharge {gpm}
- $Q_0$ = initial sprinkler discharge {gpm}
- $p$ = new pressure {psi}
- $p_0$ = initial pressure {psi}

What equation 8-1 shows is that the flow goes as the square root of the pressure (head), or the head required goes as the square of the flow. Such a relationship results in an upward curve when head is plotted in the horizontal axis and head is plotted on the vertical axis, as is normally the case with pumps.

## Friction Head

Friction head is the head lost in the delivery of the flow to the distribution device. Friction loss charts have been developed based on the Hazen-Williams equation, which is given as equation 8-3.

$$H_f = 0.2083 \times \left(\frac{100}{C}\right)^{1.852} \times \frac{Q^{1.852}}{D^{4.866}}$$

**Equation 8-3**
*Hazen-Williams equation*

where
- $H_f$ = friction head {ft; psi}
- $C$ = roughness factor of the pipe
- $Q$ = flow {gpm}
- $D$ = diameter of the pipe {in.}

Considering friction loss change from an initial condition $[H_{f_0}]$ to the final condition $[H_f]$ as flow changes, the equation is given by equation 8-4.

$$\frac{H_f}{H_{f_0}} = \left[\frac{Q}{Q_0}\right]^{1.852}$$

**Equation 8-4**
*Friction loss change*

As the system moves or zones are changed, the pipe does not change, so C and D are constant and the friction head goes as the flow to the 1.852 power. This will also result in an upward curve when plotted on the H-Q graph.

Together, friction head and pressure head result in an upward curve moving upward more than it would from either friction head or pressure head alone. Generally, the pressure head component is much larger than the friction head component.

> Consider an irrigation system including delivery and distribution pipe with 20 psi friction loss and 80 psi pressure head at a flow of 500 gpm. Figure 8-1 shows the relationship between flow and each of the head components according to equations 8-2 and 8-4. If the suction lift is 15 feet and elevation lift is 100 feet, the resulting system curve is shown in figure 8-2.

**Example 8-1**
*Friction and pressure head*

## Factors that Do Not Vary Much with Flow

### Suction Lift

Suction lift can vary but will not vary greatly. Suction lift cannot be more than 34 feet and is usually less than 15 feet. From the practical point of view, it will be assumed that suction lift does not change with flow conditions.

### Elevation Lift

Elevation lift can change as zones are changed or the system is moved (or moves) to a different elevation. Hence, changes in elevation lift should be considered.

**Figure 8-1**
*Friction and pressure head components of system curve*

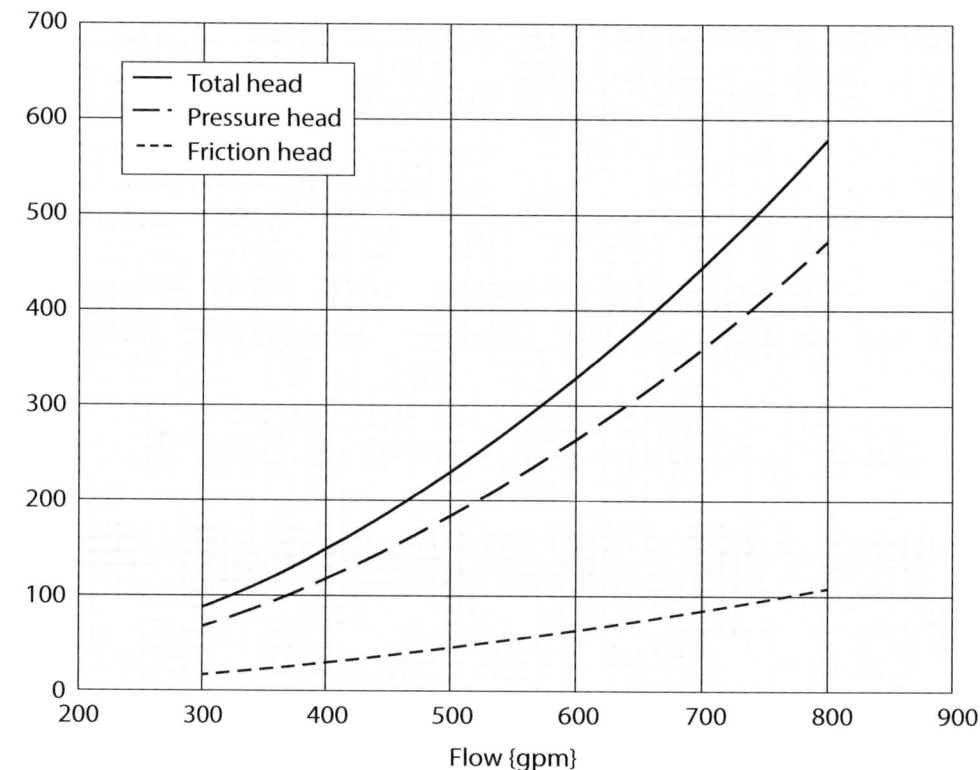

**Figure 8-2**
*Complete system curve with pump curve*

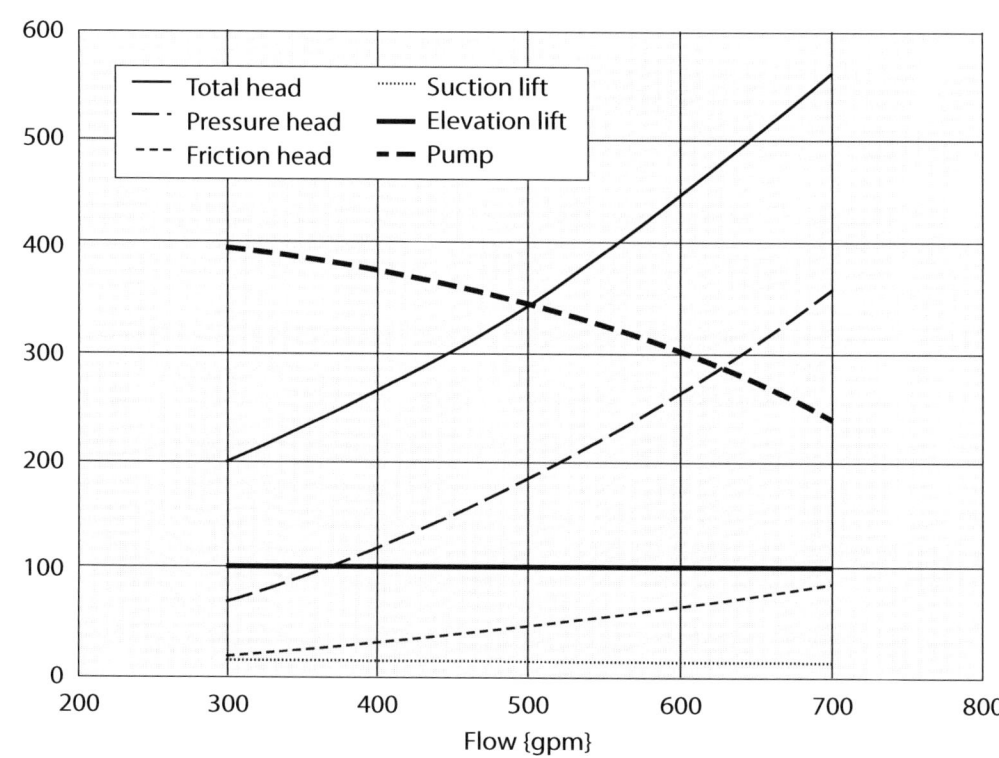

A pump was selected that exactly matched the total dynamic head for 500 gpm. At that flow, 346 feet of head is required. The suction lift is 15 feet, elevation lift is 100 feet, pressure head is 80 psi or 185 feet, friction head is 20 psi or 46 feet, and velocity head is negligible. Total head is 346 feet. The intersection of the pump curve and the total head curve occurs at 500 gpm and 346 feet of head. That is the operating point.

# Changes in the System

It is common for there to be changes in the system. The most common is when the flow in zones is not exactly the same. If the system is mobile, such as with a center pivot or traveling gun, the elevation can change. Either or both can result in a change of the system curve. In the following example, zone 2 of the system is designed to flow 550 gpm at 80 psi pressure head. This change of 10 percent of the flow results in significant differences in the pump and system performance. Figure 8-3 shows the pump and system curves for zones 1 and 2. At that scale, it is difficult to determine the values, so figure 8-3 shows the curves with expansion around the operating points.

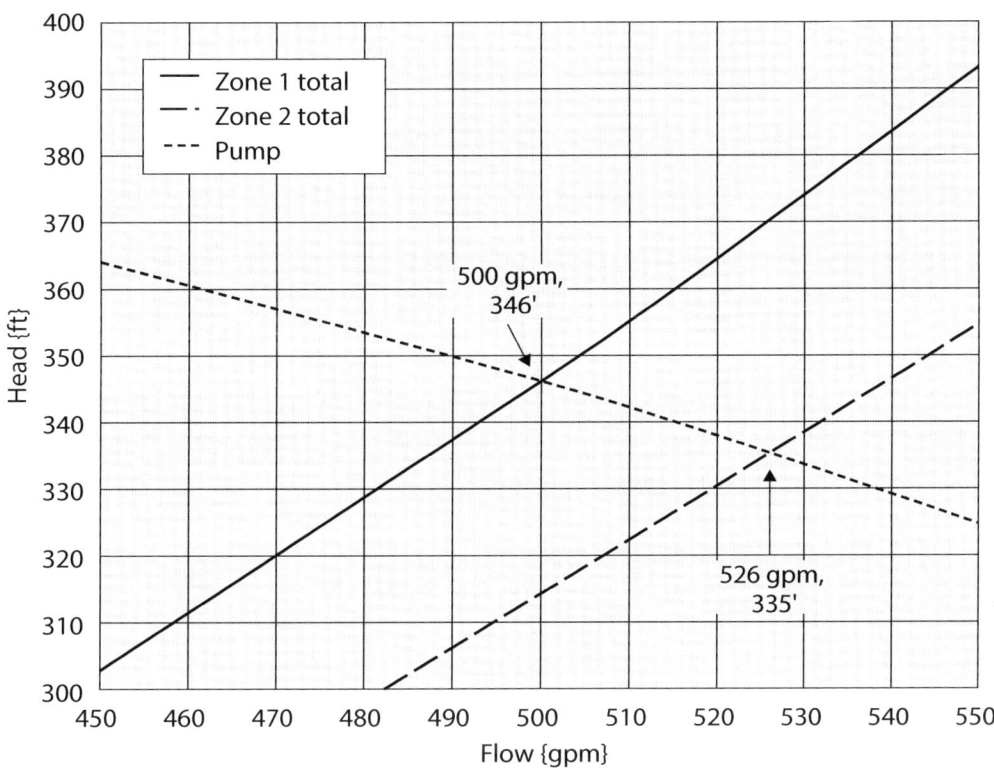

**Figure 8-3**
Expanded pump and system curves for zones 1 and 2

Figure 8-3 shows how the operating point has shifted as a result of changing from zone 1 to zone 2. While the target for zone 2 was 550 gpm at 80 psi, the result is 526 gpm at 335 feet of head. In order for the system in zone 2 to reach 550 gpm, the total head produced by the pump should be 355 feet. However, because the operating point is at 335 feet of head (20 feet or 9 psi less than design), the system in zone 2 is well below design pressure and design flow. Initially, it would appear that the system distribution is 9 psi low (355 − 335 feet = 20 feet = 9 psi), but the 355 feet of total head at 80 psi includes friction head. The operating point only determines the total head produced — not where the head demand comes from. At 526 gpm, the friction head is 4.4 feet (2 psi) less than it would be at 550 gpm, so the pressure that would have been used in friction is available for pressure head, and hence the system runs at 73 psi. Table 8-1 shows the comparison of head losses between design and actual operation.

**Table 8-1**
*Comparison of design heads and operating point heads for example system*

| System | Flow {gpm} | H$_f$ {ft} | Press {ft} | Elevation lift {ft} | Suction lift {ft} | Total head {ft} |
|---|---|---|---|---|---|---|
| Design | 550 | 55.1 | 185 | 100 | 15 | 355 |
| Operating | 526 | 50.7 | 169 | 100 | 15 | 335 |

As a result of the shift in the operating point due to the pump and system curves, the system does not operate under the design conditions. How much the operating point shifts depends on the curvature of the pump and system curves. Pressure-regulated discharge devices will limit the shift as will a flatter pump curve or a variable speed drive. Figure 8-4 shows how the system would be affected by a steeper pump curve than the one used in the example. The steeper curve results in even further reduced flow rate (now 522 gpm) and reduced head (now 331 feet) compared to design flow rate and pressure.

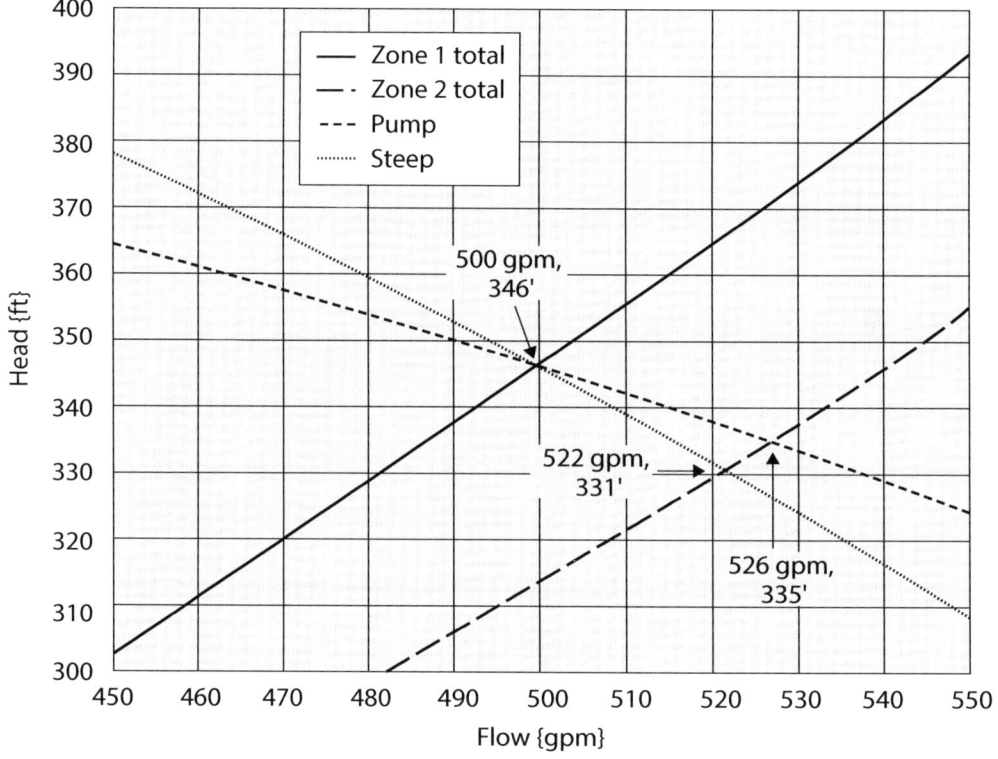

**Figure 8-4**
*Effect of a steeper pump curve*

# Practice Questions

1. The system curve includes what components that generally do not vary with flow?

   _____

   _____

   _____

   _____

2. The system curve includes what components that do vary with flow?

   _____

   _____

   _____

   _____

3. What defines where the pump will operate?

   _____

   _____

   _____

   _____

*Chapter*

# 9

# Power Units

## Learning Objectives

The following objectives are the focus of chapter 9:
- understand power plant efficiency
- compare energy sources

## Power Unit Options

The two main options for powering irrigation pumps are electric motors and internal combustion engines. In addition, a variety of fuel choices exist for internal combustion engines. The relative costs of the alternative energy sources vary greatly depending on the area where the system is located and the availability of the different fuels.

Electric motors provide the advantages of low maintenance; easy control; and quiet, efficient, convenient operation. Most motors operate at a constant speed, either 1,800 or 3,600 rpm nominal speeds (1,780 and 3,560 actual). However, variable frequency drives [VFDs] add the ability to vary motor speeds. The primary disadvantage of electric motors is the lack of three-phase electric service in some locations. In remote areas, electrical service may not be available, and it can be expensive to bring in from existing service areas. Solar and wind systems are possible options for producing electricity in some remote locations if the power required is relatively small. Even where electrical service is available, it may not include three-phase power, which is needed to power the large motors normally required for irrigation systems.

Internal combustion [IC] engines can power pumps with a variety of fuel options. They can, at the same time, run electric generators for other components of the irrigation system (e.g., automatic controllers, center pivot drive motors). An IC engine-driven pump can be installed on a riverbank many miles from the nearest power line. IC engines also provide flexible operation because the engine speed can be varied. Automatic controls are available and effective for IC engines although they are generally less common than automatic controls for electric motors.

Because of their easy and quiet operation, electric motors are the more common power unit for pumps serving turf and landscape irrigation systems. IC engines and electric motors are both common in agricultural irrigation systems. The basic advantages and disadvantages of each are summarized in table 9-1.

**Table 9-1** *Power unit advantages and disadvantages*

| Electric motors | Internal combustion engines |
|---|---|
| **Advantages**<br>Low maintenance<br>Simple to operate<br>Easy to automate<br>Longer life<br>Overall convenience | **Advantages**<br>Maintenance done by the owner<br>Flexible and mobile (in some cases)<br>Often cheaper energy costs |
| **Disadvantages**<br>Repair requires qualified electrical contractor<br>Less flexibility/mobility<br>Blackouts, brownouts<br>Safety issues: water and electricity<br>Standby/demand charges whether used or not | **Disadvantages**<br>Higher maintenance<br>Fuel prices less constant<br>Shorter life |

## *Nontypical Sources of Energy*

In addition to electricity and conventional fuels, "renewable" sources of energy may be practical in some situations, especially as alternative technologies improve and fuel costs vary. Options include wind, solar, and methane from anaerobic digestion. In addition, the energy provided by gravity should be used wherever possible.

## *Wind*

Wind power can be used indirectly for pumping by charging batteries or filling elevated tanks or reservoirs. However, wind is seldom reliable enough to directly power an irrigation system. Wind may not be available when needed, and sprinkler irrigation in windy conditions should be avoided. Typically, a cost-competitive wind turbine cannot develop the power needed for irrigation.

## *Solar*

Solar energy is more predictable than wind for most areas. In recent years the increased efficiency and reduced costs of solar panels have made them a reasonable source of energy for small-scale water pumping. Where the power requirements are low (e.g., drip or surface irrigation, livestock watering), solar energy may be a cost-effective alternative particularly where the pump site is far from a source of electricity or natural gas. If the unit needs to operate at night or during cloudy periods, batteries charged during sunny periods supply the energy.

## *Methane*

In some large-scale intensive livestock operations where irrigation is an integral component of the operation, the generation of methane from the livestock waste could be considered and the methane could be used to supply an internal combustion engine. Although the technology is still developing, anaerobic digestion systems that generate methane from livestock manure are becoming more common on farms. Methane, which is essentially natural gas, can be burned in an IC engine or an engine generator that produces electricity. In either case, the potential exists to use this fuel to power irrigation pumps, especially for irrigation systems on or near farms with digesters.

### *Gravity*

In some occasions, free gravity energy can be used when the water source is high enough in relation to the field that a pipeline delivers water at a pressure sufficient to operate the irrigation system. Gravity works particularly well where low-pressure systems (e.g., microirrigation, low pressure center pivots) are used at an elevation significantly lower than the water source. Even if there is not enough elevation drop to completely eliminate the need for a pump, any elevation drop will reduce the power required for the same system with no elevation drop.

## Internal Combustion Engines

Most IC engines used in irrigation are either modified spark ignition automotive engines or diesel engines. The power required for irrigation pumping differs from automotive service in the following ways:

- Internal combustion engines on pumping plants often use natural gas or propane. These fuels have less lubricity and require special lubricating oils and valves.
- Irrigation engines operate unattended.
- Irrigation engines operate at continuous speeds, under full load.
- Irrigation engines operate tremendously long nonstop hours compared to automotive service. There is little opportunity for cooling during the irrigation season. A system that runs 1,200 hours per year is equivalent to driving a car about 72,000 miles per year.
- A major overhaul typically is required at about 6,000 hours.
- Irrigation engines should last at least 10,000 hours before being replaced.
- Diesel engines, which initiate combustion by compression rather than a spark, have similar characteristics except that their life span and time between major overhauls is much longer.

Once the pump has been selected, several factors must be considered before selecting an internal combustion engine. First, the IC engine must have adequate net continuous power for the application. In addition, other factors can lessen the net available continuous power. Some power loss will occur in the power transmission system (transmission or gear head); some power will be used by engine cooling fans unless a heat exchanger is used; and the net power must be corrected for altitude, air temperature, and other engine-powered components such as an alternator. Often, IC engines are rated on maximum intermittent power and not on net continuous power, so the designer must be aware of the difference.

Engine power is affected by a number of factors including the equipment and accessories installed on the engine, climatic conditions, elevation, and engine wear. Because a normal radiator fan uses a large amount of power, irrigators may use heat exchangers rather than a radiator fan to increase the net remaining power available at the drive shaft of the engine.

The net continuous brake horsepower available for pumping is estimated by derating the engine; that is, adjusting the power to allow for air temperature, elevation, and speed of operation and also deducting the power used by additional engine components (fan, muffler, generator, etc.). Engine horsepower decreases about 1 percent for every 10°F above 85°F. Engine horsepower also decreases about 1.5 percent for every 500 feet above sea level. Turbocharged engines are not affected as much by temperature and elevation as are naturally aspirated engines.

## *Natural Gas*

Where readily available, natural gas can be a relatively low-cost and reliable fuel for irrigation engines. Historically, most natural gas engines used were industrialized versions of automotive engines such as Chrysler, Ford, International, and Oldsmobile. As automobile manufacturers turned away from large displacement, high horsepower engines in their automobiles, fewer appropriate automotive engines have been available. As a result, more industrial natural gas engines such as Cummins, Caterpillar, Waukesha, and others are being used more in irrigation applications. The price of these engines is significantly higher than automotive engines, but their life span is correspondingly longer. There have also been modifications to some diesel engines to allow them to operate on a mixture of approximately 85 percent natural gas and 15 percent diesel fuel.

The life expectancy of spark ignition engines is less than electric motors or diesel engines, and the cost of maintenance for them can be significant.

## *Liquefied Petroleum Gases*

In some areas irrigation engines use liquefied petroleum gases [LPG], propane, or butane (or a mixture of both). LPG has the advantage of being portable, so LPG may be a good choice if a pump station has to move from one water source to another. Pumping units are available that will run on either LPG or natural gas. Some irrigators buy an LPG engine for use when portability is needed, or they use LPG temporarily until the natural gas line is installed.

## *Diesel*

Diesel fuel is also a common irrigation pumping energy source particularly for large systems. The popularity of diesel is due mainly to the easy availability of diesel fuel and large (140–200 horsepower and larger) diesel engines in all areas of the country. Also, as systems get larger, irrigators put more hours on their pumping units. The added life of a diesel engine pays off in these instances. Many farms have mechanics able to maintain diesel engines, which are common for farm equipment, so they can apply that expertise to a diesel pumping unit as well. Diesel engines tend to require less maintenance throughout their lives, and they last significantly more hours than spark ignition engines.

### *Tractor Power Take-Off Drives*

Tractors with power take-off [PTO] pumps are sometimes used in irrigation applications. PTOs are usually driven by either diesel- or gasoline-powered tractors. They may be a reasonable low-cost option for periodic duty if a tractor is available for this application during the irrigation season.

## Electric Motors

Many irrigation pumping plants use electric motors. They have a long expected life, require minimal maintenance, and are very reliable (see fig. 9-1). Electric motors lend themselves easily to automation. Electric motors are available in a wide range of sizes, and their maintenance is considerably less than any other power unit. Convenience is a main advantage of electricity. Many irrigators choose an electric motor because it runs at the push of a button.

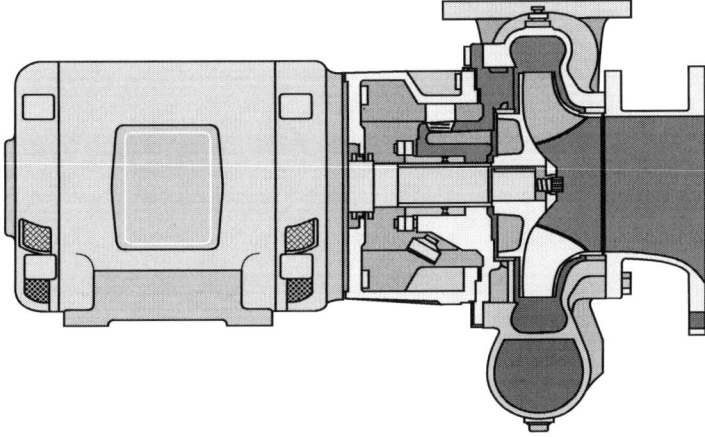

**Figure 9-1**
*Electric motor close-coupled to an end-suction irrigation pump*

Electric motors are used widely where three-phase power is readily available. A disadvantage to using large-horsepower electric motors is the high cost of supplying three-phase power to the pump site if it is not available nearby. Single-phase motors are not practical when the power requirements exceed 10 horsepower. For motors between 10 and 40 horsepower, it may be practical to buy a three-phase motor and use a phase converter to convert single-phase supply to three-phase where three-phase power is not available. Before installing such a system, the electricity supplier should be consulted to ensure that the single-phase power service line can support the size of the motor(s) being installed.

Electric motors operate at high and relatively constant efficiency when under normal load. Motor efficiency does not change substantially with age. However, as table 9-2 shows, there is a slight decrease in motor efficiencies when the load on the motor decreases from 100 percent to 50 percent of the rated motor power. Motor efficiency decreases substantially when the load is less than 50 percent of the rating power of the motor. Larger motors are generally more efficient than smaller motors.

When choosing a motor, it should be remembered that the energy consumed (kilowatt hours [kWh]) is a function of the load, not the power shown on the nameplate of the motor. Motors draw power in proportion to the load. If the applied load is less (or more) than the motor rated power, the motor uses less (or more) energy than it would under full load. For example, if the load were 100 horsepower, a 200-horsepower motor would be 90 percent efficient (50 percent loading, see table 9-2), while a 100-horsepower motor would also be 90 percent efficient (100 percent loading). Thus, they would both consume the same amount of energy per hour.

**Table 9-2** Typical standard motor efficiencies

| Motor Horsepower | Motor efficiency {%} at 100% Load | Motor efficiency {%} at 50% Load | Motor Horsepower | Motor efficiency {%} at 100% Load | Motor efficiency {%} at 50% Load |
|---|---|---|---|---|---|
| 3 | 84 | 81 | 60 | 90 | 87 |
| 5 | 85 | 80 | 75 | 90 | 90 |
| 7.5 | 86 | 81 | 100 | 90 | 89 |
| 10 | 87 | 83 | 125 | 91 | 90 |
| 15 | 89 | 85 | 150 | 91 | 90 |
| 20 | 89 | 85 | 200 | 92 | 90 |
| 30 | 89 | 83 | 250 | 92 | 91 |
| 40 | 90 | 89 | 300 | 92 | 91 |
| 50 | 90 | 89 | 350 | 92 | 91 |
|  |  |  | 400 | 92 | 91 |

## Types of Motors

Many different types of motors can be used in irrigation applications, depending on the conditions under which they are expected to operate and whether they will be exposed to the elements. Open drip-proof motors are common and the least expensive. They are constructed so that water cannot enter from the top, but they are exposed to the elements in some areas. Totally enclosed fan-cooled [TEFC] motors are better suited to exposure to the elements but are more expensive and may be subject to heating problems if not installed correctly. In rare cases, explosion-proof motors may need to be considered. Motors are also available with various "soft-start" options so that their start-up current is less. These may need to be considered in rural areas where the line servicing the pump site is limited in the current it can supply. Any questions regarding these alternatives should be directed to a qualified electrician familiar with the needs of large horsepower motors in your service area.

## Motor Selection

When selecting an electric motor, the following points should be considered:

1. Using the standard pump power equation (equation 4-3), calculate the horsepower required by the pump. The required power should not be read off the pump curve. However, the power curves on a pump curve can be a useful check on the calculations.

2. Consider the future. What will happen when the nozzles wear? Is there a possibility that more acres will be irrigated? Will the golf course be expanded? Will a corner system be added to a center pivot? The difference in the capital

cost between the minimum-sized motor and the next larger one is often small.

3. Consider the voltage and speed requirements (see fig. 9-2, 230 or 460 volts, 1,760 rpm). The utility company determines voltage availability. There may also be limitations to the motor size because of the electrical service or existing switchgear.

4. Consider the application relative to the type of motor required:
    - *open drip-proof or TEFC* — In special situations an explosion-proof motor may be required.
    - *vertical or horizontal shaft* — Vertical solid-shaft or vertical hollow-shaft motors are available depending on the application where a vertically mounted pump is to be used.

5. Electric motors can produce the nameplate power rating marked on the motor (see fig. 9-2, 40 horsepower). Unlike internal combustion engines, derating is not required.

6. Because the larger motor of two choices can perform the job with greater ease, it will run cooler and be less susceptible to overload shutdowns. These situations can occur when the ambient temperatures are high or the system flow changes. As mentioned previously, a 20 percent larger motor (e.g., 60 horsepower vs. 50 horsepower) will *not* use 20 percent more energy for the same application. The extra size, if the cost difference is small, may be a wise investment. However, before making that decision, an electrical supplier should be consulted to see if the demand charge is based on actual demand or nameplate rating.

7. The higher the motor horsepower, the greater the cost of related switch gear and wiring. Some comparative shopping is needed. In some cases, the cost difference is small (e.g., between 60 and 75 horsepower). In other cases, the cost difference is substantial (e.g., between 100 and 125 horsepower).

8. Electric motors have a service factor [SF] listed on the nameplate (see fig. 9-2, 1.15 SF). The SF represents a safety factor built into the design of the motor. For example, open drip-proof motors usually have an SF of 1.15 (115 percent). This number means that under continuous load, these motors can develop 1.15 times the horsepower shown on the nameplate. Therefore, a 150-horsepower open drip-proof motor could theoretically be used in applications requiring up to 172.5 brake horsepower, but this is not recommended. TEFC motors often have an SF of 1.00 (no extra capacity). In most cases, the SF should be left as reserve capacity. However, in some rare situations, operating slightly into the SF may be warranted (e.g., using a 150-horsepower open drip-proof motor with a 1.15 SF to develop 155 brake horsepower). It is important to consider the long-term costs and returns when making the decision regarding motor size. If a larger motor increases reliability, it must be considered. It is always an option to re-evaluate the system design to determine whether any changes can be made to the system design or the pump efficiency to reduce the required brake horsepower.

9. Consider motor efficiencies. Energy-efficient motors are sometimes used in irrigation applications. The efficiency of these motors is 2–4 percent higher than the standard motors shown in table 9-2. Energy-efficient motors are more expensive, but some utility companies provide incentives to promote the use of high-efficiency motors. The premium that is paid for a higher efficiency motor produces a return every year. The extra cost of high-efficiency motors may be relatively small, depending on the number produced. Energy-efficient motors are particularly appropriate for new pumping plant installations, for replacing older motors, or for systems using a large number of hours per year. In other circumstances, the potential savings in energy cost should be balanced against the extra cost of the energy-efficient motor. Current trends in the industry, partly because of increased energy costs and the increased availability of energy-efficient motors, are toward using high-efficiency motors in irrigation systems.

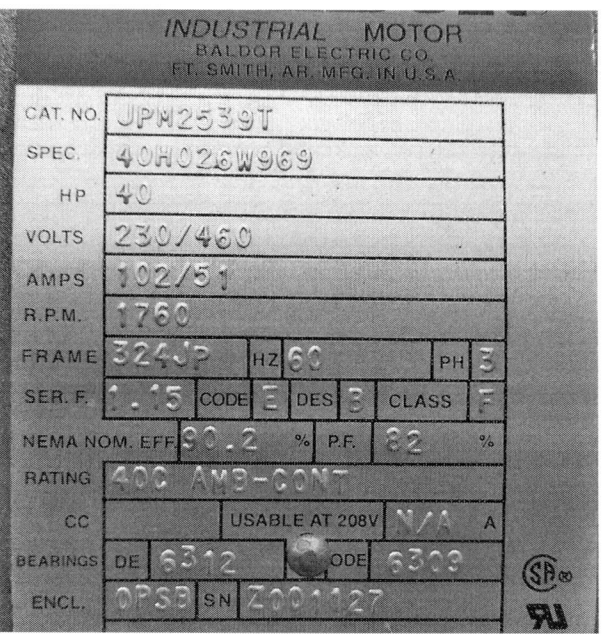

**Figure 9-2**
*Typical electric motor nameplate (40 horsepower; 230/460 volts; 102/51 amps; 1,760 rpm; 60 hertz; 3 phase, 1.15 SF)*

## Electrical Metering, Demand Charges, and Rate Choices

Electric power rates must be sufficient to pay for the electric power supplier's capital investment and operating costs, which include cost of the energy consumed. For year-round and relatively uniform customer loads, the two factors (capital and operating) may be combined in a rate structure that charges only for the amount of energy used. For intermittent or seasonal loads, such as irrigation, electric power suppliers usually impose two charges. To cover operating costs, there is the regular energy charge; and to cover fixed costs, there is either a hookup or demand charge, which is based on the motor power regardless of the hours pumped per year. This structure reflects that the utility company's capital investment costs are a function of the maximum total load/demand {kW or hp} connected to the system, regardless of the total energy consumed. Rate structures vary widely among electric power suppliers and geographic regions. How the electricity is produced can also significantly affect the rate charged per kilowatt-hour. If the irrigation system runs many hours during the irrigation season, a lower energy charge is more important than a lower demand charge. If the hours of pumping are low and the connected power is high, the converse is true.

## *Controlling Demand Charges*

When there is a demand charge, it is usually structured as either a specific charge per connected nameplate horespower (or kilowatt) or a stated charge per maximum measured kilowatt of demand. When the demand charge is based on the measured kilowatt, a demand meter is used. These meters measure kilowatt-hour on one register or set of dials and record the maximum kilowatts of demand elsewhere on the meter. The maximum kilowatts for each billing cycle, as recorded by the demand meter, is the highest average use of power over a 15- or 30-minute period during the billing cycle. The time period varies with different power suppliers.

To minimize the demand charge, an irrigator should manage the system to keep demand as consistently steady as possible. With most irrigation systems, the greatest demand usually occurs when starting the system and/or filling the pipelines. Therefore, systems should always be started slowly and the demand meter reading noted to avoid very short-term, high demand that applies to the entire billing period. If a large system is started at the end of one billing cycle and continues into the start of the next cycle, that large demand will apply to both billing cycles. In this case, it may be advantageous to delay the start of the system until the new billing cycle begins.

## *Time-of-Use Electrical Rates*

Time-of-use [TOU] electrical rates encourage pumping at times when the normal daily demand for electricity is not at a peak. As an incentive, irrigators are charged a lower rate per kilowatt-hour to irrigate during off-peak hours. The peak-use period typically ranges between 11 a.m. and 7 p.m. (sometimes only on weekdays) with other times considered off-peak. The particular utility company sets the exact peak and off-peak hours. Some TOU rate schedules include more than two rate structures (i.e., low, medium, and high rates). In areas of the country where summer and winter electrical demand is significantly different, the winter rates may vary from the summer rates whether or not TOU rates are in effect.

As much as possible, irrigation systems with motor-driven pumps should be run during off-peak hours to take advantage of TOU cost savings. Fortunately typical off-peak hours (night and early mornings) are also good times to irrigate from an irrigation efficiency standpoint. Golf courses in particular are well suited to irrigating at these times. However, if this strategy reduces the time period for irrigation, the shorter run times must be taken into account. For example, irrigating for 18 hours per day rather than 24 increases the required pumping flow rate by 33 percent but does not reduce the total volume of water to be pumped. Turf and landscape irrigation designers (e.g., golf courses) are familiar with irrigating less than 24 hours per day, but this concept is relatively uncommon in agricultural irrigation systems.

Some irrigation pumping systems can be designed to use both an electric motor and an IC engine. If this is possible, and the energy cost for the IC engine is significantly less than the peak energy charge for electricity, an irrigator can use the IC engine during peak electrical demand and switch to an electric motor during the off-peak times. Turbine pumps can be equipped with a dual input gear head (electric motor on top and a horizontal driveshaft for the IC engine on the side) to give the irrigator this choice.

Another way to take advantage of lower TOU electrical rates is possible if the system has access to a water storage reservoir at an elevation much higher than the main water source and pumping unit. When rates are low, the pump can be used to fill the reservoir. Then when rates are high, rather than pumping with electrical energy, water can be diverted back down from the reservoir and gravity used to pressurize the system.

## Power Plant Efficiency

There are many types of power plants available for pumping. Generally they are categorized by the energy source. The Nebraska Performance Standards for Irrigation Pumping Plants gives criteria for different power sources. Pumps are assumed to be 75 percent efficient, and electric motors are assumed to be 88 percent efficient. Table 9-3 represents the expected water horsepower hours per unit of energy for five energy sources. It is possible to achieve higher pump efficiencies and electric motor efficiencies than those assumed.

**Table 9-3**
*Nebraska Pumping Plant Performance Criteria [NPPPC]*

| Energy source | Energy unit | Bhp-h/unit[a] | Whp-h[b]/unit[c] |
|---|---|---|---|
| Electric | Kilowatt-hour | 1.18[e] | 0.885 |
| Diesel | Gallon | 16.6 | 12.5[d] |
| Natural gas | 1,000 cubic feet | 88.9[g] | 66.7 |
| Propane | Gallon | 9.2 | 6.89 |
| Gasoline[f] | Gallon | 11.5 | 8.66 |

[a] Horsepower hours {Bhp-h} is the work accomplished by the power unit including drive losses.
[b] Water horsepower hours {Whp-h} is the work produced by the pumping plant per unit of energy at the NPPPC.
[c] The NPPPC are based on 75 percent pump efficiency.
[d] Criteria for diesel revised in 1981 to 12.5 Whp-h/gal.
[e] Assumes 88 percent electric motor efficiency.
[f] Taken from Test D of Nebraska Tractor Test Reports. Drive losses are accounted for in the data. Assumes no cooling fan.
[g] Manufacturers' data corrected for 5 percent gear-head drive loss and no cooling fan. Assumes natural gas energy content of 1,000 Btu per cubic foot.

## Comparing Energy Sources

The NPPPC can be used to compare energy sources. An in-depth comparison is shown in chapter 14 of *Irrigation, Sixth Edition*. Further analyses of investment alternatives are shown in chapter 15. The following example compares assumed costs for three power sources. Assumed costs are electricity at $0.10 per kilowatt-hour, diesel fuel at $2.60 per gallon, and natural gas at $8.00 per 1,000 cubic feet.

|  | NPPPC Whp-h/unit | $/unit | $/Whp-h |
|---|---|---|---|
| Electricity | 0.885 | 0.10 | 0.113 |
| Diesel fuel | 12.5 | 2.60 | 0.208 |
| Natural gas | 66.7 | 8.00 | 0.120 |

**Table 9-4**
*Energy source cost comparison*

With the assumed prices, this shows that electricity has the cheapest operating cost. It does not account for initial cost or the expected life of the unit.

# Practice Questions

1. Name two advantages of electric motors for pumps.

    (1.) _____

    (2.) _____

2. Name two advantages of internal combustion engines for pumps.

    (1.) _____

    (2.) _____

3. The NPPPC allows comparison of what important pump parameter?

    _____

    _____

    _____

    _____

# Chapter 10

# Automation and Control

## Learning Objectives

The following objectives are the focus of chapter 10:
- understand pumping system controls
- understand how pumps can be automated
- introduce variable frequency drives

## Pumping System Controls

There are basically three ways to control a pumping system. The easiest, but likely not the least expensive in the long run, is to design the pump system to meet the maximum pressure and flow and add control valves, which can regulate (reduce) the pressure or flow, control start-up rate, or other specific demands. These types of controls use a valve to dissipate (throw away) the energy in the water that isn't needed. As a result, they may be simple and easy to install, but are the least efficient of the options. The most common of these is the throttling valve.

A second method is to control the power unit and take advantage of the affinity laws and pump curve to produce the desired result. If the power unit is an internal combustion engine (diesel, gasoline, natural gas, or propane) the control is by adjusting the fuel flow to the engine, which in turn affects the rotational speed. If the power unit is alternating current electric, control is achieved by changing the frequency of the input voltage/current and thereby changing the rotational speed of the pump. This is a variable frequency drive [VFD]. In direct current systems, control can be achieved by regulating the voltage to the motor.

A third method is on/off control. An example is a sump pump with a float where the pump turns on at an upper limit and turns off at a lower limit. This results in the pump cycling on and off. Another method involves a pressure header or tank with on/off control. If properly designed, the pump does not cycle much, and the on/off control produces acceptable results. An example is a municipal system that pumps to a water tower. Individuals have access to the line pressurized by the elevation of the tank and can cycle units as needed. Each individual uses a much smaller portion of flow than the pump and tank system is capable of, so the pump cycles infrequently and pressure is relatively constant.

For many years, on/off control has been used for pressure and/or flow limits. Usually the issue is low pressure (perhaps due to a line break) or low flow (indicating an upstream flow blockage). A sensor is wired into the system and when a threshold is reached, the power or fuel to the power unit is interrupted.

# Automation

With VFDs or other controls, it is possible to greatly reduce the need for operator intervention at the pump site. The pump site can be automated in a variety of ways to provide a variety of automated responses. The most common way pump sites are automated is to provide pressure control so the discharge pressure is constant regardless of the flow in the system. With some modifications pressure control mechanisms can perform other automated tasks.

## Pressure Control

The simplest, and often least expensive, way to control pressure is by installing an automatic pressure-reducing valve on the discharge line. These valves are usually hydraulically controlled diaphragm valves with pilot piping systems (small-diameter piping and controls) that control the flow of water to the top of the diaphragm and maintain a constant (adjustable) pressure on the downstream side of the valve (see fig. 10-1). As long as the pump is producing more than the desired pressure, the valve reduces the pressure to that required by the system. Some energy is lost as a result of the pressure drop across the valve.

**Figure 10-1**
*Automatic diaphragm valve*

When the valve is modulating the pressure, the pressure pilot senses the downstream pressure and if it is too high, the pilot closes slightly and forces water onto the diaphragm. This causes the valve to close slightly to reduce the downstream pressure (see fig. 10-2). When the pressure drops to the set point, flow to the diaphragm stops, and the valve maintains its position and downstream pressure.

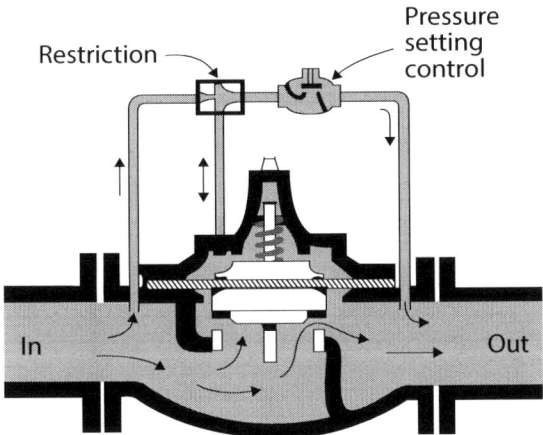

**Figure 10-2**
*Automatic pressure-reducing valve in operation*

For IC engines, an automatic throttle control may be added to the engine and connected to a discharge pressure gauge with set points. This is similar to cruise control in an automobile. When the discharge pressure gets too high (or low) the throttle control adjusts to slow (or speed up) the engine to maintain the pressure between the set points. For electric motors, VFDs perform the same function as an automatic throttle control for IC engines.

For computer or programmable controller-based control systems, the automated pressure set point may be a function of the flow rate in the system. For example, for a center pivot sprinkler system with a corner-arm attachment, a control algorithm can be developed that increases the pump discharge pressure as the arm swings out and the flow rate increases. Another way to obtain suitable results is by installing pressure or flow regulators on each sprinkler. This works in the same way as pressure control valves, but there is one at each sprinkler. As long as the pressure at the regulator is more than the regulator pressure, constant pressure on the sprinkler is maintained and flow is constant.

## Other Automatic Control Valves

The same valves that are used as pressure-reducing valves may be equipped with different pilot control piping to perform a variety of other automated tasks (see fig. 10-3).

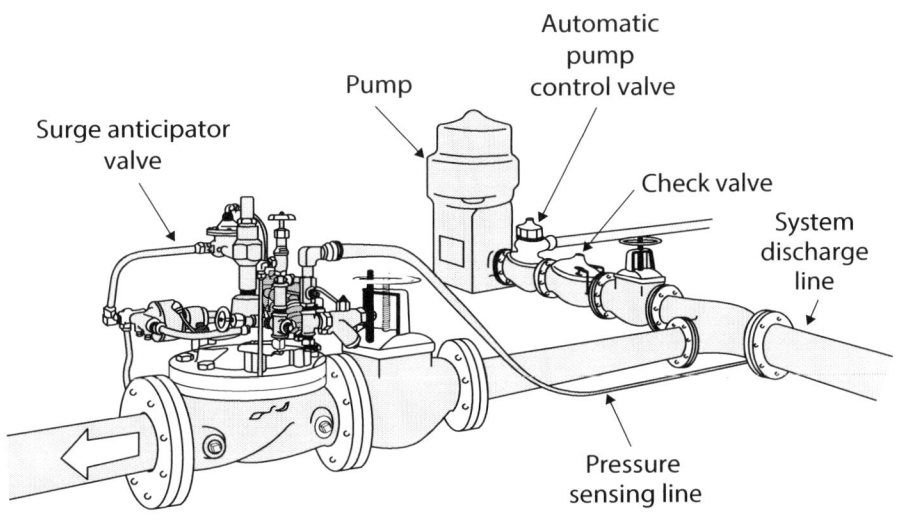

**Figure 10-3**
*Various automatic control valves*

**Chapter 10: Automation and Control**

## Pressure Relief Valves

When relief valves sense a pressure higher than the set point, their pilot system enables them to open quickly and discharge water away from the system into a sump or back to the water source (see fig. 10-4) so that the system does not experience excessive pressures. These valves must be installed on the side of the main line to discharge away from the system. They are not installed as in-line valves. These valves work well to eliminate the gradual buildup of excessive pressure, but they do not react fast enough to prevent damage due to water hammer.

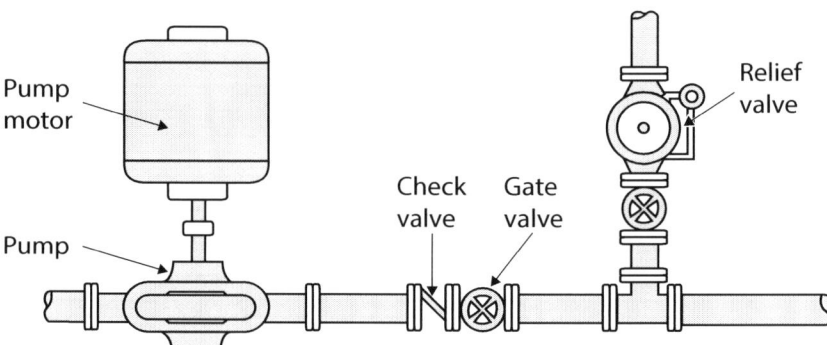

**Figure 10-4**
*Automatic pressure relief valve*

## Surge Anticipation Valves

These valves have a pilot system that senses a sudden drop in pressure (as when the pump shuts down) and opens quickly before the wave of water flows backward. The valves discharge water away from the system into a sump or back to the water source (see fig. 10-3). In this way the pump station is spared the force of the water hammer caused by the returning water. These valves are especially suited to installations where the pump station is below a long uphill main line.

## Pump Control Valves

These valves reduce the pressure that pumps start against and also moderate the pressure changes in the system as pumps start and stop. Pump control valves are usually used where a number of pumps are installed in parallel at a common pump site or where the pump must start against a filled pipeline. Like pressure relief valves, they are mounted on the side and recycle water to the water source (see fig. 10-3). When the pump starts, the valve is in the fully open position. The pump starts against no backpressure because the water just circulates back to the source. The valve closes gradually and the pressure builds slowly. When the pressure reaches the pressure in the system (on the downstream side of the check valve), the check valve opens and the pump starts pumping into the main line. The start-up valve continues to close and when it closes completely, the new pump is pumping its full flow into the system. The pump does not start against pressure, and the system does not experience a sudden rise in pressure because the pressure and flow added by the new pump are gradually added to the system.

When the pump shuts down, the same procedure is automatically followed in reverse. The valve slowly opens, allowing water to start to flow back to the source; the pressure and flow being added by the pump slowly decrease until the check valve closes and all of the pump's flow is back to the source. When the valve is fully open the pump motor shuts off. Again, there are no sudden pressure surges in the system. These valves are particularly important where vertical turbine pumps with long columns are connected in parallel. VFDs can be installed on one pump in multiple pump systems to perform a similar function.

### *Combination Valves*

In some situations, a single diaphragm valve is equipped with combination pilot systems to perform two or more functions. A common example is an electric solenoid shutoff (it opens and closes according to an electric signal), pressure-reducing combination valve. When the electric signal causes the valve to open, it acts as a pressure-reducing valve. When the signal directs the valve to close, it closes, shutting down the irrigation system. A more complicated valve, used only in unique situations, could have a pilot system that results in a pressure-reducing, surge-control, pressure-sensitive opening solenoid control valve.

The complications of designing and maintaining these sophisticated pilot systems make them relatively uncommon in typical irrigation systems. Plugging of the pilot system can be a problem because the valve is controlled by the water being pumped, which can often contain silt or sand. The pilot piping can be equipped with a fine cartridge filter, but if the water contains a large amount of sediment, the filter will plug quickly.

## Integrating Pump Stations with Irrigation Controllers

The simplest method of integrating a pump with an irrigation controller is by the use of a pump start relay. The controller generally operates on 24 volts alternating current and is not capable of powering a pump, but it is capable of powering a relay in the 120 volts alternating current (or higher) power circuit of the pump. The controller tells the relay when to start and when to stop. Electric motors can be further controlled with variable frequency drive technology.

## Variable Frequency Drives

Electric motors used for irrigation pumping usually operate at a nominal 1,800 or 3,600 rpm. This constant pump speed limits the operational flexibility and/or energy efficiency of the pumping system. VFDs allow the speed of the motor to be changed by changing the frequency of the alternating current supplying the motor. As the frequency decreases, the motor speed also decreases. This decrease in motor speed reduces the pressure and flow to match the requirements of the irrigation system. Normally, it also reduces the energy consumed.

Typically, pumps and power units are selected to meet the maximum flow and pressure requirements. However, during a typical irrigation cycle, the operating conditions for some irrigation systems (e.g., golf courses) vary considerably depending on which zones or sets are being irrigated at any one time. Under these circumstances, the required flow rate can decrease as different areas are irrigated. In some instances, the flow is throttled by partially closing a discharge valve. This adjustment increases the friction loss, reducing the flow and pressure in the irrigation piping. Throttling decreases the power demand and energy consumption of the motor (because the flow rate is reduced), but it also causes energy to be lost across the valve resulting in wasted energy and higher than necessary energy costs. For example, if the valve is throttled to cause a 10-psi drop in a system that is pumping 1,000 gpm with an 85 percent pump efficiency and a 90 percent motor efficiency, the energy loss would be equal to 5.7 kilowatt-hours per hour of wasted energy.

If the pump always produces more pressure than necessary, the simplest and most cost-effective way to correct the situation is to trim the impeller to a smaller diameter. The amount of trimming required can be calculated using the pump affinity laws. This trimming will reduce the pressure produced and reduce the energy consumed.

If the pumping unit must provide water over a wide range of flow rates, trimming the impeller does not work because this is a permanent and irreversible change. In these cases, multiple pumps can be used and automated. Alternatively, a VFD control could be added to the pump motor to allow the pump to operate effectively over a wider range of flows and pressures. Even with pump stations having multiple pumps, a VFD can be connected to one pump motor to improve the performance of the overall station and reduce the pressure surges when pumps start up or shut down.

Adjusting the pump speed may not yield the same total head and flow rate obtained under throttled conditions. Where VFDs are used, they are commonly integrated into an automated control system to provide constant pressure and/or flow as required.

Some observers have raised concerns about the effect of variable speed drives on motor temperature because an increase in temperature can shorten the life of an electric motor. A report published by the Electric Power Research Institute [EPRI] confirms that VFDs can raise motor temperatures 3–15 percent but adds that VFDs of more recent design produce less than a 5 percent temperature increase.

A VFD can also affect pump efficiency. Regardless of how a pump is powered, its performance is still dictated by its performance curve, and the pump operates somewhere on the curve for the speed at which it is actually turning. If a pump is slowed and the flow rate changes such that the operating point is at an inefficient part of the performance curve, nothing can be done with the power unit to change that. Proponents of VFDs often make the mistake of assuming that if the pump is 85 percent efficient at its design point, it will be 85 percent efficient at all operating points when the VFD changes the speed. This assumption rarely holds true. If the speed changes are so small that the pump efficiency does not change, the changes are probably so small that a VFD is not justified. In some cases, if the pump has been selected correctly to operate efficiently over the complete range of operating points, the best efficiency may actually be encountered when the speed is something less than maximum. Therefore, to ensure that the best pump is chosen for the application, the irrigator should know in advance whether a VFD will be used. If a VFD is simply added to a pump that was selected assuming a constant pump speed, the improvements will be less than those possible by selecting the pump knowing beforehand that it will be operating at various speeds.

As far as motor efficiency is concerned, the lower the speed, the less efficient will be the motor and the VFD. The VFD itself also is not 100 percent efficient. It consumes some power although the loss is minimal. Down to about 50 percent of the maximum speed, the motor/VFD efficiency may decrease only slightly, but at speeds lower than 50 percent the efficiency falls dramatically. The EPRI report estimates that at 50 percent speed the VFD efficiency is about 84 percent compared to 95 percent at full speed. At 25 percent of maximum speed, the drive efficiency falls to 53 percent.

VFDs, like all electrical equipment, must be protected from adverse environmental conditions including dampness and extremes in temperature and altitude. VFDs should be operated in an atmosphere free of corrosive gas and dust.

## VFD Potential Problems

### Operational Problems

#### *Insulation Stress*

A VFD operates by converting alternating currenct [ac] voltage into direct current [dc] voltage (both positive and negative) and then sending out a "chopped" dc voltage in varying lengths of time through pulse width modulation or PWM. By controlling the pulse width and timing, the motor sees what appears to be an ac voltage at a frequency controlled by the PWM. However, the voltage rises very quickly and isn't a true ac voltage. The quick rise in the voltage causes voltage to be induced in the motor windings, which gets reflected back into the motor cables. These voltages can be several times the presumed voltage of the system. There may also be corona effects which produce ozone. Both overvoltage and ozone tend to break down insulation.

#### *Harmonic Distortion*

Because of the rapidly changing voltage, secondary waves can be produced. These waves, when in synchronization with the main wave, produce voltage overshoots. When exactly out of synchronization, they produce voltage undershoots. The result is variance from the desired smooth ac voltage. Harmonic distortion can also result in an imbalance among the three phases. Harmonic distortion for one phase is shown in figure 10-5.

**Figure 10-5**
*Harmonic distortion of voltage*

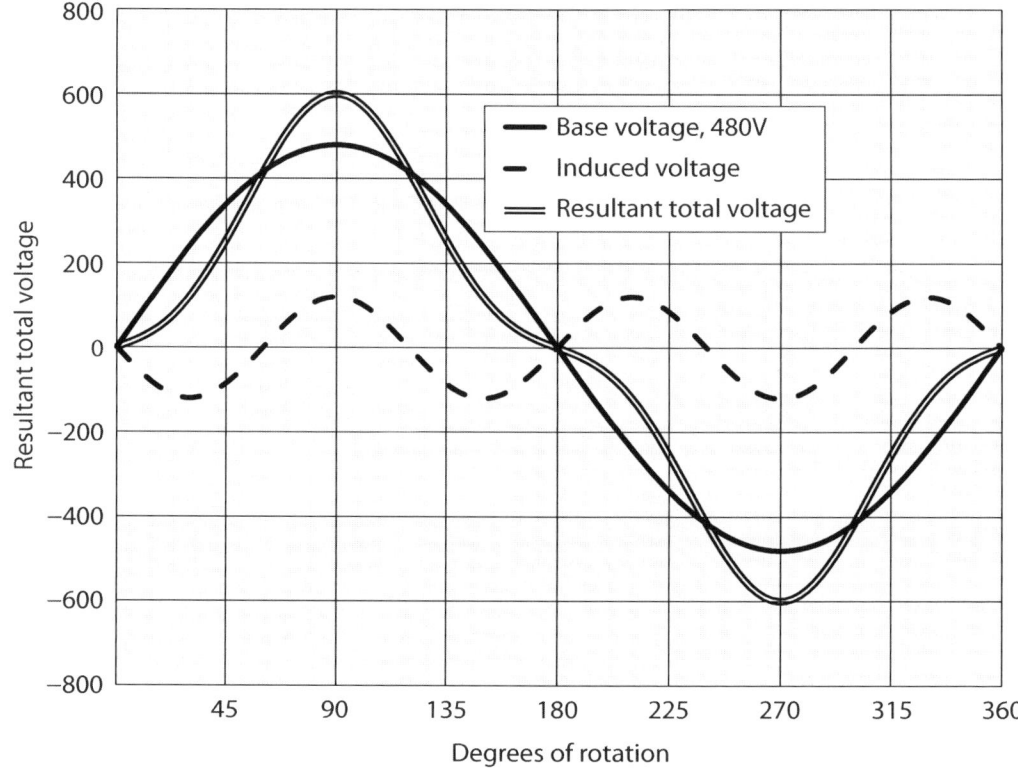

## Bearing Currents

The distortion of voltage and imbalance can result in current seeking ground. One path to ground is through the motor bearings, and these currents can cause electro-discharge machining [EDM] to occur resulting in pits in the bearing race resulting in premature bearing failure.

## Resonant Frequency

Similar to the harmonic distortion, a resonant frequency can occur with the motor causing it to vibrate excessively. If the VFD tries to run the motor at a resonant frequency, it can result in catastrophic destruction of the motor and pump assembly.

There are solutions to most of the potential operational problems, but they must be installed.

According to the U.S. Department of Energy Motor Systems Tip Sheet #14,

> *Electronic adjustable speed drives, known as variable frequency drives (VFD), used to be marketed as "usable with any standard motor." However, premature failures of motor insulation systems began to occur as fast-switching, pulse-width-modulated (PWM) VFDs were introduced. The switching rates of modern power semiconductors can lead to voltage overshoots. These voltage spikes can rapidly damage a motor's insulation system, resulting in premature motor failure. Insulation problems can be addressed by using an "inverter duty" motor.*

*Inverter-duty motors are wound with voltage spike-resistant insulation systems. Some use inverter-grade magnet wire to minimize the adverse effects of waveforms produced by VFDs. Other designs are wound with adjacent coils that are separated to minimize voltage potential. Improved insulation systems reduce degradation of motors that are subjected to transient voltage spikes. A greater thickness or buildup of premium varnish (through multiple dips and bakes) minimizes the potential for internal voids, and a motor with a lower heat rise design has increased resistance to voltage stresses. Manufacturing quality also affects the corona inception voltage (CIV), the point at which partial electrical discharges occur because of ionization of air around the conductor. CIV is a measure of the ability of a motor's windings to withstand voltage stresses.*

## Control Problems

Fundamental to the proper operation of a VFD is its ability to choose a pump speed that matches the desired operating conditions. In some cases that may not be possible. Consider whether a VFD would solve the problem presented in chapter 8 involving two zones of a system as depicted in figure 8-3. Assuming that the pump was initially operating at 1,775 rpm producing 500 gpm at 346 feet of head, is it possible to change the speed and produce 550 gpm at 346 feet of head? The answer to the question is not straight-forward. It involves developing a best match pump curve and seeing how it matches with the system curve. The new speed cannot be directly calculated using the affinity laws because the target is a curve (the system curve), not a single point. Using the affinity laws to develop a new pump curve and looking at the intersection of the pump at the new speed and the zone 2 system curve, the flow objective of 550 gpm is matched at 1,836 rpm. If the VFD were programmed to hold constant flow, it would eventually settle at 1,836 rpm (within the accuracy of the flow meter[1] and the VFD). However, the head is 355 feet. Similarly, if the head for zone 2 is matched at 346 feet with 1,809 rpm, the flow is 539 gpm. Similarly, if the VFD were programmed to hold constant head, it would settle at 1,809 rpm[2]. These two conditions are shown in figures 10-6 and 10-7. A compromise might be met at 1,822 rpm producing 541 gpm at 351 feet of head. The point of this process has been to show that it is not likely that both the flow and head objectives can be met exactly with a VFD when the system curve changes, and the ability to match conditions depends upon both the system curve and the pump curve[3]. While the flow and head conditions weren't exactly met, they were very close.

### Caution Notes

[1] Pressure transducers used in VFD pump systems are probably more accurate that flow meters. Typical flow meter accuracy is about 2.0 percent at best.

[2] Pressure transducers are likely to be 0.4 percent or better.

[3] It is much more difficult (or perhaps impossible) to meet new zone conditions with a flat pump curve.

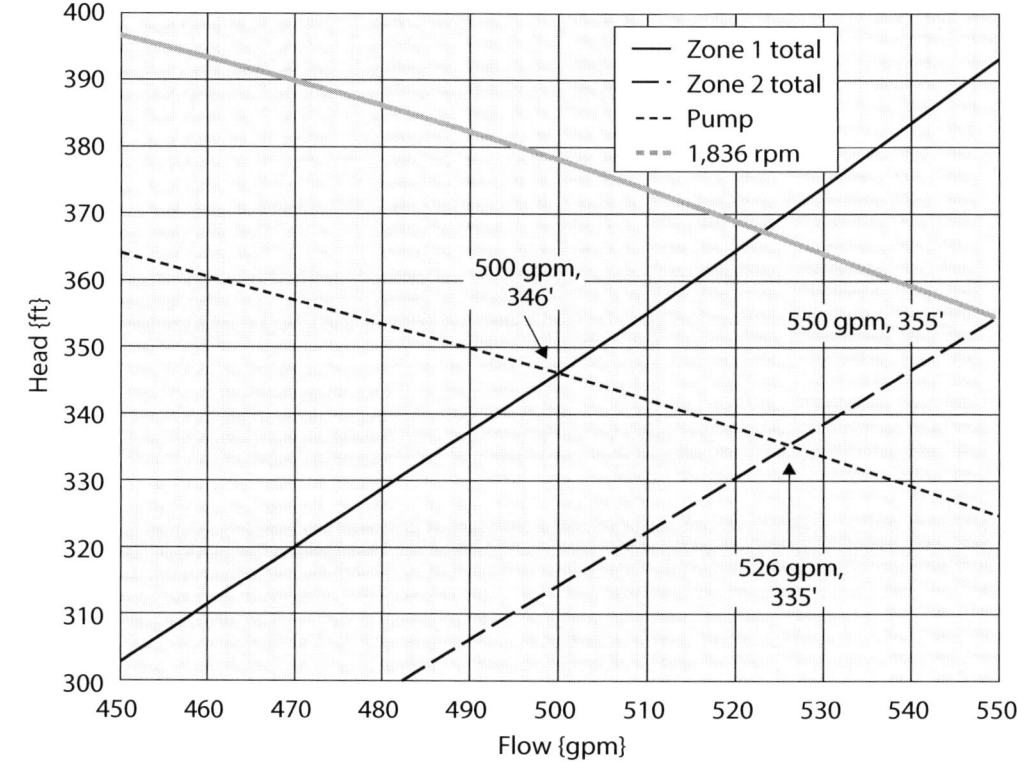

**Figure 10-6**
Speed matched to desired flow

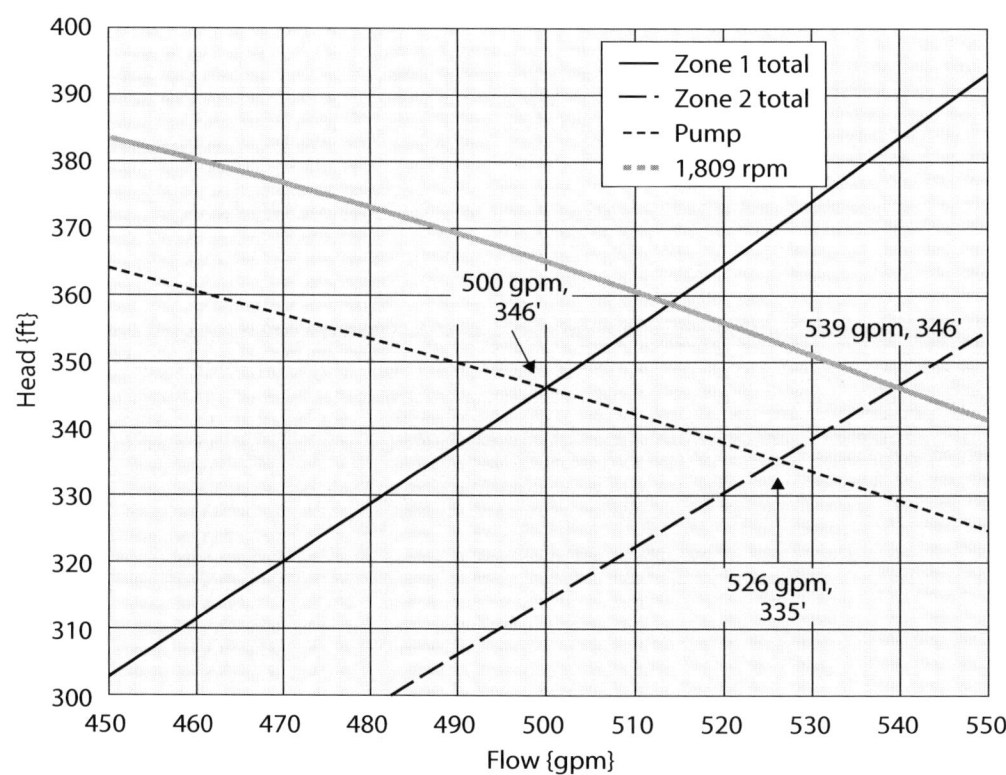

**Figure 10-7**
Speed matched to desired pressure

## Other VFD Issues

There are other issues that arise as a result of using VFDs. One is the effect on torque and horsepower with speed changes. Figure 10-8 shows the effect of motor speed on horsepower and torque.

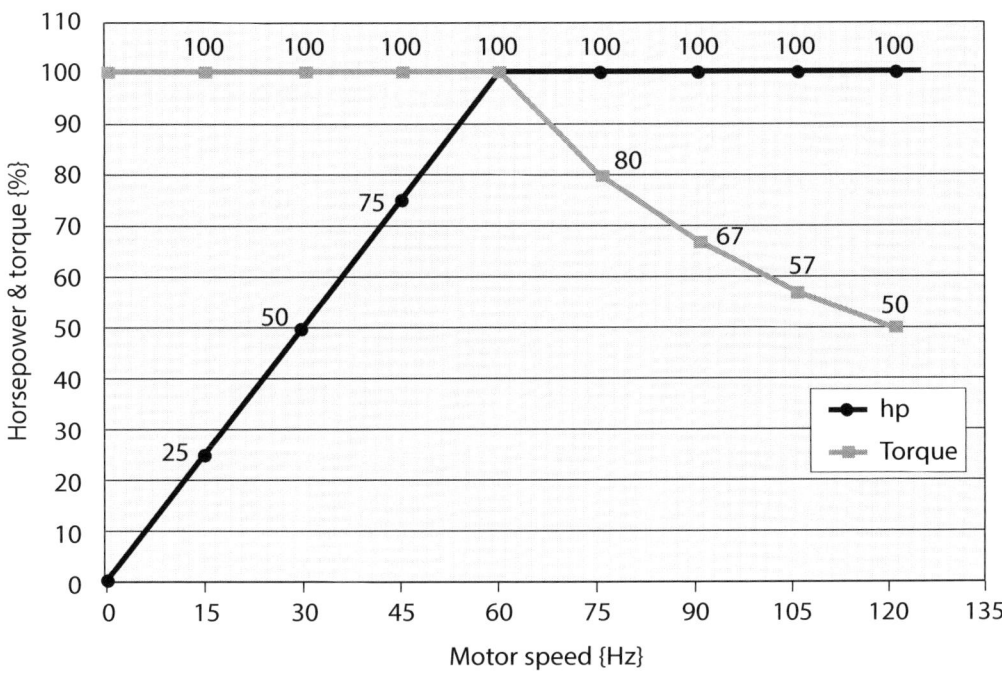

**Figure 10-8**
Torque and horsepower changes with speed change (after Joe Evans)

Fortunately, as speed is reduced, pump horsepower is reduced by the cube, so reducing speed to ½ reduces horsepower to ⅛. However, as speed is increased, pump horsepower still goes up by the cube, but motor power does not. A 15 percent increase in motor speed results in no more motor horsepower but a 52 percent increase in pump horsepower. Hence, if an increase in speed is anticipated, the motor must be appropriately oversized. An excellent resource to consider when using a VFD for pump control can be found on the Irrigation Association website under RESOURCES/Technical Resources/NRCS on Variable Speed Drives for Irrigation Pumps. This reference also includes construction specifications.

# Practice Questions

1. Name three common ways of controlling pumps.

    (1.) _____

    (2.) _____

    (3.) _____

2. How do VFDs work?

    _____

    _____

    _____

    _____

3. What are the four commonly encountered problems with VFDs?

    (1.) _____

    (2.) _____

    (3.) _____

    (4.) _____

4. Which pump curve is least likely to benefit from a VFD in flow control (flat or steep)?

    _____

    _____

    _____

    _____

# Chapter 11

# Operation and Maintenance

## Learning Objectives

The following objectives are the focus of chapter 11:
- understand how VFDs can lower pumping costs and improve operation of the system
- know important maintenance issues

## Using VFDs to Lower Energy Costs

VFDs eliminate the energy otherwise wasted by throttling the pump discharge valves on constant speed pumping units. The economic benefit of VFDs depends on the decrease in horsepower, the portion of operating time at the reduced speed, the electrical energy costs, and the capital cost of the VFD. One potential benefit of a VFD may be savings in annual energy cost. The additional cost of the VFD is the annualized capital cost of it, which is partly determined by the life of the drive. Conditions that favor VFDs are significant reductions in horsepower and a significant portion of the total operating time at the reduced speed. However, if the portion of time at the reduced speed is very high, it may be more cost-effective to trim the impeller on the pump and live with a slightly reduced capacity for the small portion of the time that more pressure is required.

To accurately determine if the addition of a VFD is cost-effective, the following need to be known:

- increased capital cost of the VFD
- expected life span of the VFD
- total annual energy costs without the VFD
- hours of operation at various flow/pressure combinations
- estimate of the total annual energy costs with the VFD
- rate that energy costs will rise in the future
- rate of inflation (time value of money)

# Using VFDs to Improve the Operation of the Pumping Station

Some users have found that the energy savings from VFDs are much less than they expected but, at the same time, the improvements to the overall operation of the system still justify the cost of the VFDs. Such improvements include less operator intervention and delivery of water at a more constant pressure. Operators of large turf or landscape and golf course irrigation systems find this feature especially attractive. This situation is particularly true where irrigation districts have added VFD drives to one or more pumps on their large pump stations delivering to a variety of end users. The pressure fluctuations in the system are smoothed greatly, and the pressure surges due to pumps starting and stopping can be almost eliminated. These benefits are realized as long as the various pumps, the VFD, and the control system are designed carefully. Generally, VFDs are better suited to systems with many hours of use and flow rates that vary significantly. The capital cost of a VFD is the same whether the pump operates 100 hours per year or 1,000 hours per year, but the energy savings are a function of the hours pumped. A hidden benefit of a VFD is the reduction in pressure cycling. Pressure cycling can significantly reduce the life of the piping system, especially the fittings. A detailed discussion can be found on the IA website under RESOURCES/Technical Resources/Designing, Operating, and Maintaining Piping Systems Using PVC.

The decision to use a VFD depends greatly on the ability to correctly identify the load profile (e.g., table 5-1). VFD manufacturers and suppliers can use that information and other energy cost and operating cost data to estimate if the cost of the VFD is offset by the energy savings over time. Another important consideration is power quality. VFDs are sensitive to transient voltages and voltage fluctuations, which can be common in rural areas. The harmonic current caused by VFDs can also produce additional heating in electric motors. An expert always should be consulted before a decision is made to install a VFD.

# Prepackaged Pump Stations with VFD Control

Prepackaged pump stations are available for a wide variety of pumping applications. These stations are available in standard configurations for various ranges of varying flow and pressure and usually incorporate pressure and/or flow control using automatic control valves alone or in conjunction with VFD controls. The construction and operational convenience of these packages make them a common choice for golf courses and other systems with widely varying flows.

# Motor Management and Maintenance

One of the attractive features of electric motors is their reliability compared to internal combustion engines. Owners of IC engines are almost always familiar with maintenance requirements. Although motors need much less maintenance, the type of maintenance they need is less understood by the average operator. The simplicity of a motor leads some operators to mistakenly believe that no maintenance is required. With a small amount of regular maintenance, an electric motor should last from 20,000 to 40,000 hours.

# Overheating

Because motors aren't 100 percent efficient, some of the energy is converted to heat. Motors are designed to operate within specified temperatures. Most irrigation motors are designed for temperatures up to 70°F above ambient. Too much heat greatly reduces motor life by damaging the insulation of the motor coil windings. The following can cause excessive heat:

- poor ventilation, preventing adequate air circulation through the motor
- low voltage, causing the motor to draw excessive current
- voltage imbalance among the three phases of a three-phase motor
- possible increased load on the irrigation motor over time (e.g., if the sprinkler nozzles wear and the flow rate increases)

The following steps are recommended to extend motor life:
- maintain a good flow of air around the motor (keep ventilation openings clear)
- operate the motor at the rated power or less
- keep voltage imbalance among phases at 4 percent or less
- assess the system if a motor frequently cuts out (Flows and/or pressures may have changed significantly since the system was first installed. An irrigation specialist or pump testing agency can help calculate the present load on the motor.)
- protect the switchgear from the elements unless it is designed for outdoor installations (A tarp is not adequate protection.)
- protect irrigation motors with a lightning arrester
- provide power surge protection wherever the utility power is subject to spiking

An additional cause for motor failure is overloading. An electric motor will attempt to pump the amount of water demanded by the system. The motor will draw as much electricity as required to drive the load it is coupled to. If it draws more electricity than it was designed for, heat will build up in the motor and cause damage or the motor will stop. To reduce capital costs of an installation, it is tempting to purchase a motor that is slightly undersized for the job. The difference in annual cost between a motor that will be slightly overloaded and the next larger size, when spread over the acres irrigated and a 20-year period, may be very small.

### Outdoor Operation

Pump motors should be located in a dry, well-ventilated location. Water from the pump packing gland should be drained away from the motor. Motors operating above a pool of water present a serious electrical safety hazard.

Most irrigation motors are of an open drip-proof design. They have large openings in the frame that allow air circulation for cooling but are located so that water falling from above does not enter the motor. In the off-season, mice or other small animals may enter these holes and chew on the insulation inside the motor. Wire mesh should be screwed over these openings before the motor is installed. Most pump houses do not stop mice from getting inside the pump house. Open drip-proof should not be installed outdoors. The openings are potential entrances for rain, snow, and dirt that can damage the motor.

If a motor is installed outside, it should be a totally enclosed fan-cooled motor. Because these motors are more expensive, it may be more cost-effective to purchase an open drip-proof motor and build a pump house.

## Preseason and Postseason Checklists

In addition to measures listed above, the following preseason and postseason steps are recommended:

- Switch boxes should be checked in the spring.
- All the connections in the panel should be tightened and the switchblades cleaned if required.
- The electrical cable from the panel to the motor should be checked for cracks and poor connections. If any of the connections appear burnt, an electrician should be called.
- There should be no openings in the switch boxes where mice can enter.
- In dry climates, the insulation on the windings tends to dry and crack. When a motor is about 10 years old, it is advisable to have the winding insulation recoated. The bearings can also be economically replaced during this process to ensure dependability.
- On motors with oil reservoirs, the oil should be checked for evidence of moisture or oxidation. If the oil appears discolored, it should be drained and replaced.
- In the off-season, oil-lubricated motors should be rotated once a month to maintain a coating of oil on the bearings.

# Routine Inspection and Maintenance[1]

## Inspecting the Pump

A typical inspection includes looking at and walking around the pump. Regular inspections help develop a sense of what the pump should sound and feel like. Feel the motors and pumps. Are there any strange noises or vibrations? Can you detect a bearing or motor that is unusually hot? Is there a new odor or electrical smell? Use caution around drive couplings and electric controls.

Many pump breakdowns trace back to the stuffing box. A badly leaking packing gland or mechanical seal will cause problems. Water spraying into a motor or bearing frame will infiltrate the pump end bearing. It will wash all lubrication from the bearing, causing rust and imminent failure.

If water collects under a horizontally mounted motor, the ventilation fan (which blows onto the motor winding) will pull the water into the motor. This may cause a burned-out motor. Water squirting up into a vertical, hollow-shaft motor of a vertical turbine pump will cause the same problems. These motors are *not* water-cooled. Electric motor service life depends on a dry, clean atmosphere. Elevate a horizontally mounted pump at least 6 inches off the floor, and install a line to drain the leakage away from the motor.

Vertical turbines have drain connections in the bottom of the discharge head. Connect these to a hose and drain the head. Keep the drains open and flowing.

The packing gland and grease cup are easily adjusted to the manufacturer's instructions. Special lubrication for the grease cup and packing is available from local suppliers. The grease cup lubricates the packing and aids in priming horizontal pumps. When adding a packing ring, be sure it's new and clean. Carefully align the gland without cocking it. Tighten evenly to achieve the manufacturer's specified leakage. This means minimum leakage with a cool stuffing box. Replace dried and worn packing that has lost its lubrication. This requires a special tool called a packing hook. Packing hooks are also available from local suppliers.

After removing all the packing, inspect the shaft sleeve. If the sleeve is grooved or worn, packing replacement will have a short life. Replace the sleeve. This requires disassembling the pump. Horizontal units usually must go back to the shop. Vertical turbines usually require motor removal and head shaft renewal.

If the pump is equipped with a mechanical seal, never allow it to run dry, even for a few seconds. Water lubricates the seal faces. A dry run will burn it out. At the first sign of a leak, replace the seal. This will require disassembly, which a pump technician normally does. Seals are easily damaged by improper handling, so carefully follow manufacturer's instructions.

---

[1] This section based on Hydraulics & Pump Seminar Book by Cornell Pump Company

Pumping sand and silt will naturally shorten the life of the packing, sleeve, seals and wear rings. Good planning and site selection can ensure maximum service life.

If the pump is in the shop for a sleeve replacement, measure the wear on the wear rings. If the wear is 1/32 of an inch or 0.030 per side, restore the clearances with new wear rings and impeller repair..

## Lubrication

If the motor has Zerk grease fittings, it requires greasing. Some of the smaller sizes, usually 3 to 5 hp pumps, won't have Zerk fittings and don't require greasing.

When adding grease, be sure the grease and the fittings are absolutely clean. The code number for the proper grease is EP-2. Other greases, such as multi-purpose types, may work, but bearing manufacturers recommend only EP-2. The exception is if the motor or pump manufacturer specifically recommends a different lubricant.

To lubricate electric motor bearings, remove the relief grease plug. Using a hand grease gun, pump the new grease into the fitting until it shows at the drain. Do this when the unit is not running so you avoid getting grease into the motor. Leave the drain plug out for a few days to let the excess grease work its way through the drain, not into the motor.

The bearings will run unusually hot for about 20 minutes after greasing because the bearing is purging the grease from the balls and race. As the bearing warms up, it turns the grease to oil. It's this mist of oil that actually lubricates the bearing. That is why it is essential to use the Code EP-2 for proper melting temperature.

Pumps mounted on bearing frames (those that have a separate motor) are normally greased through the bearing cover. Excess grease accumulates in the large cavity of the frame. It takes years to fill the frame. Follow the manufacturer's instructions in the operator's manual for greasing frequency. A drain plug is usually a pipe plug near the bottom of the frame.

Proper motor ventilation is just as critical as lubrication. The temperature of the motor winding determines its life. Normal temperature means a long life.

Many motors have rodent screens installed on the vents. These keep critters out, but they require periodic cleaning. Keep them free of lint, chaff, weeds, dirt and other debris to ensure a free, cool, air flow.

Well ventilated shelters help protect pumping equipment and switch gear from sun and rain. The sun's direct rays can add 10 to 20 degrees of ambient temperature to the motor temperature. For every 18° Fahrenheit temperature rise above the motor nameplate rating, the expected motor life is reduced by one-half. Thermostat controlled exhaust fans help keep the inside temperature and air flow cool in pump houses.

## Vibrations

An extreme vibration could be the result of a misaligned drive coupling or the start of bearing failure. Some pump units can actually twist on their bases if the base construction is too light or if they are not secured and grouted properly to the foundation.

Pipeline misalignment can also lead to vibration. Unsupported pipelines full of water put a tremendous weight load on the pump casing. Pipelines can break the casings if the weight load is severe enough. The pipelines must be supported so the pump can be removed with no stress or strain on it. It is a good practice to use one flexible-type pipeline coupling in either the inlet line or discharge line.

A noise developing in a pump that has otherwise been running quietly usually indicates a bearing is beginning to fail. Replace the bearing immediately. Neglect could irreparably damage the motor or the frame. A bearing that repeatedly fails indicates a possible misalignment or strain. Occasionally the wrong type of bearing is installed, or it is not heavy enough for the application. If in doubt, request a B-10 bearing life calculation from the pump manufacturer.

## The Electric System

Electric switch gear needs periodic inspection and maintenance as well as the pump and motor. This requires an electrician who is experienced in controls and pump starters. He should check the contacts in the starter and replace any that show signs of uneven or heavy pitting. If neglected, these are going to heat and cause high current to trip out the overload protection device.

Overload heaters should have the proper rating and the overload trip rating should be properly adjusted. The electrician should check and tighten every screw in the panel. After several years, normal heat and temperature changes tend to loosen the terminal screws. A loose connection will cause heat, burn out wiring, damage the contactor and/or cause short motor cycling and overheating. Remember, maintaining a low temperature rise in the electric motor will ensure a long service life.

# Practice Questions

1. How can a VFD reduce pumping costs?

   _____

   _____

   _____

   _____

2. Vibrations can lead to what kinds of problems in pumps?

   _____

   _____

   _____

   _____

3. Why is overheating an issue in electric motors?

   _____

   _____

   _____

   _____

# Chapter 12

# Irrigation Pumping Costs

## Learning Objectives

The following objectives are the focus of chapter 12:
- understand fixed costs
- understand variable costs
- know how to use time value of money to compare systems

## Comparing Alternatives

When comparing different alternatives, it is necessary to differentiate between fixed (annual) costs and variable (operating or hourly) costs of pumping. Fixed costs do not change regardless of how much the pump is used. Variable costs are proportional to the number of hours that the pump operates. Generally, if a pump runs for a small number of hours, an energy source with low fixed costs is an advantage. With a large number of pumping hours per year, a system with low variable costs is advantageous.

## Fixed Costs

Fixed costs apply to pumps and power units and their associated annual costs.

### Pumps

Almost all costs associated with the pump itself are fixed costs. Fixed pump costs include the following:

- *capital replacement cost* — This is the capital cost of the pump amortized over the life of the pump. A less expensive pump can have a higher capital cost per year if it has a shorter life span. If the hours used per year affect the life of the pump, this cost would actually be a combination fixed/variable cost.
- *maintenance costs* — Most pump maintenance items are done once or twice per year (before start-up and at the end of the irrigation season) regardless of how much the pump runs. If the hours of operation per year affect the pump maintenance, this cost would actually be a combination fixed/variable cost.

### Power Units

Power units have significant fixed and variable costs, which vary with the type of power unit. For power units, fixed costs include the following:

- *capital replacement cost* — This is the capital cost of the power unit, amortized over the life of the power unit. Electric motors generally have lower initial costs and a longer life than IC engines, so their annual capital replacement cost is significantly less. However, the capital cost of bringing in three-phase power must also be included in the fixed costs. The same situation occurs if the irrigator is responsible for part of the capital cost of supplying a natural gas line to a natural gas power unit. These charges vary widely depending on how the particular utility company recovers these capital costs. Natural gas and LPG engines are generally less expensive than diesel units but have a shorter life span, so they may have a higher annualized capital cost. If the hours used per year affect the life of the power unit, this cost would actually be a combination fixed/variable cost.
- *maintenance costs* — Maintenance costs are fixed if they are done at regular intervals regardless of the use (e.g., electric motors). For IC engines most maintenance is a variable cost performed after a set number of operating hours. A well-maintained IC engine lasts much longer than a poorly maintained unit, thus reducing annual fixed costs for the engine.

## Demand or Other Fixed Energy Charges

If there is a service charge that is related to demand, irrespective of the number of hours pumped, it is a fixed cost. Also, if there is a monthly meter, fuel tank rental, or minimum charge, irrespective of use, this is also a fixed cost.

There are other fixed costs (e.g., insurance) associated with the ownership of an irrigation system, but unless these costs vary depending on the type of pumping unit selected, they do not need to be considered when choosing a pumping system.

## Variable Costs

Variable costs include energy, maintenance, and some pumping costs.

### Pumps

Few costs associated with the pump itself are variable. Only those maintenance items that are performed after a defined number of hours are variable costs. These items include packing gland, belt or bearing replacement, and well servicing.

## Power Units

Power units have significant variable costs including the following:

- *variable energy costs* — The cost paid per unit of energy consumed (gal, kWh, MCF, etc.) is the main variable cost of the pumping system. Typical rates of energy consumption for different energy types are given in chapter 9 (NPPPC criteria) to predict the energy cost before choosing which power unit to use.
- *maintenance costs* — Maintenance that occurs at defined hourly intervals represents a variable cost. IC engines have significant maintenance variable costs including oil changes and tune-ups. A normal irrigation pump engine operating continuously at full load is equivalent to driving a car about 10,000 miles per week.

# Operating Cost

Operating costs are those directly related to the operation of the equipment. Some operating costs, such as repairs, may increase with the amount of use per season or with the speed at which the equipment is run, but they are generally considered to be proportional to the amount of time the equipment is run.

## Energy Costs

Energy costs are a major portion of operating costs. Energy for pumping and pressurizing the system depends on both the amount of pressure the pump must produce (total dynamic head) and the amount of water pumped. The amount of water pumped depends on crops, soils, weather, and system efficiency. Overall pumping energy consumption depends not only on the system but on the efficiency of the pump and the efficiency of the engine or motor. Figure 12-1 shows the relationship among the key variables of energy cost.

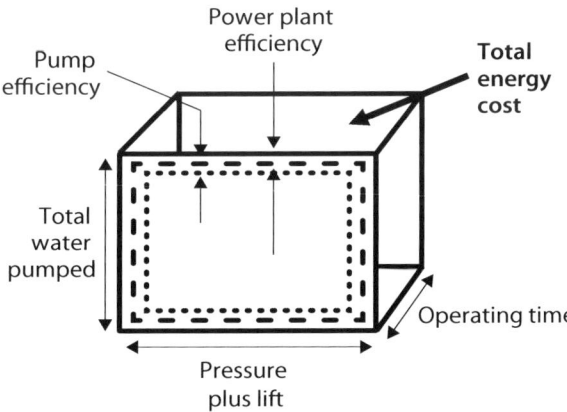

**Figure 12-1**
*Total energy cost*

The dotted inner box represents the water power needed. It is the product of the total water pumped and the pressure plus lift. The ratio of the size between the dotted box and the dashed box is the pump efficiency, and the ratio of the size between the dashed line box and the solid line box is the power plant efficiency. This is to show that each of the components contributes to the total water power needed. The cost of energy is determined by the size of the solid line box and the time the system runs, which produces the three-dimensional, black-bordered box.

Several factors determine the overall pumping (energy) cost. The purpose of irrigation is to deliver usable water to the plant (net irrigation, $IR_{net}$). Irrigation application efficiency [$E_a$] is the ratio of net irrigation water [$IR_{net}$] to gross irrigation water [$IR_{gross}$ or Q], as shown in equation 12-1.

**Equation 12-1**
*Irrigation application efficiency*

$$IR_{net} = IR_{gross} \times E_a$$

Irrigation efficiency depends on system design, soils, operation, and management.

The second factor of energy cost is the efficiency of the pumping plant. Typically irrigation pumps convert 70–80 percent of the power delivered to the pump into the product of pressure and flow, but much lower efficiencies are encountered. Initial pumping plant efficiencies depend on pump design and selection of the pump, but wear can significantly reduce the efficiency of a pump. Water power is proportional to gross irrigation and total head. Brake horsepower [Bhp] is water power [Whp] divided by pump efficiency [E], and power input [$P_{IN}$] is brake power divided by the efficiency of the power supply unit [$E_{pp}$].

Equation 4-3 is used to calculate the brake horsepower, but does not account for the efficiency of the power unit ($E_{pp}$). Equation 12-2 incorporates $E_{pp}$ to determine the input power needed.

**Equation 12-2**
*Power input*

$$P_{IN} = \frac{Bhp}{\left(\frac{E_{pp}}{100}\right)}$$

where
- $P_{IN}$ = input power; power supplied to the power unit {hp}
- Bhp = power discharged from the power unit to the pump {hp}
- $E_{pp}$ = efficiency of the power unit {%}

Ultimately, the input power is controlled by the net amount of irrigation needed [$IR_{net}$] and the irrigation efficiency [$E_a$], the pump efficiency [$E_p$], and the power unit efficiency [$E_{pp}$].

Electric energy is purchased by the kilowatt-hour, and diesel energy is purchased by the gallon. In both cases, the energy is purchased by the basic energy unit, not by any unit of power delivered to the water. Hence, the efficiencies above are all factors in total energy cost, so the need to maximize efficiencies is self-evident. Typical irrigation efficiencies run from 35 to 70 percent for surface systems, 60 to 85 percent for sprinkler systems, and 85 to 95 percent for microirrigation systems. Diesel engines are typically 30–35 percent efficient, and electric motors are 80–93 percent efficient depending upon the size and design. It must be kept in mind, however, that the electricity may have been produced by a fossil-fueled power plant operating in the 35–45 percent efficiency range. Comparing the very lowest efficiency combinations with the very highest, one sees that there is a difference of a factor of almost 4. The lowest is with $E_a = 0.35$, $E_p = 0.70$, and $E_{pp} = 0.30$, giving overall efficiency of 0.074. The highest is $(0.93) \times (0.80) \times (0.35) = 0.260$. The ratio is $0.260 \div 0.074 = 3.51$. It does not necessarily mean the most energy efficient system is the best, but there are significant energy cost differences. Furthermore, if there is a high lift or high pressure, the significance of the individual efficiencies is magnified. Using a microirrigation system operating at 20 psi with a 10-foot pumping lift and the highest efficiencies indicated above as the baseline, the following are the relative diesel pumping costs.

Figure 12-1 and table 12-1 show that total energy costs are subject to inefficiencies in irrigation, pumping plants, and power plants. Inefficiencies are made much worse with high pumping lifts, excessive friction losses, and/or operating pressure. In table 12-1, increases in irrigation, pumping plant, and engine efficiencies drop relative energy costs by 62 percent. A decrease in the operating pressure (without decreasing irrigation efficiency) almost negates all the difference between the high pressure sprinkler, high lift, and low efficiency; to the medium pressure sprinkler, high lift, and high efficiency. Efficiencies and pressures are factors the irrigator can control. Often the irrigator cannot control lift.

|  | Efficiencies | | | Lift | | Pressure | | Total head | | Comparative | |
| --- | --- | --- | --- | --- | --- | --- | --- | --- | --- | --- | --- |
|  | Irrigation | Pump | Diesel | {ft} | {m} | {psi} | kPa} | {ft} | {m} | Factor | Costs |
| Micro, low lift, high E | 0.9 | 0.8 | 0.35 | 10 | 3 | 20 | 138 | 56 | 17 | 223 | US$ 1.00 |
| Micro, high lift, low E | 0.9 | 0.7 | 0.30 | 500 | 152 | 20 | 138 | 546 | 166 | 2,890 | US$12.97 |
| High spr, low lift, high E | 0.9 | 0.8 | 0.35 | 10 | 3 | 85 | 586 | 206 | 63 | 818 | US$ 3.67 |
| High spr, high lift, low E | 0.5 | 0.7 | 0.30 | 500 | 152 | 85 | 586 | 696 | 212 | 6,629 | US$29.76 |
| High spr, high lift, high E | 0.9 | 0.8 | 0.35 | 500 | 152 | 85 | 586 | 696 | 212 | 2,762 | US$12.40 |
| Med spr, high lift, high E | 0.9 | 0.8 | 0.35 | 500 | 152 | 45 | 310 | 604 | 184 | 2396 | US$10.76 |

**Table 12-1**
*Comparative energy costs*

## Repair and Maintenance Costs

Repair and maintenance [R&M] costs are usually taken as a fraction of the initial cost of the equipment. For motors, the cost might be as high as 2 percent of the initial cost per year. Movable systems or systems subject to damage from tractor related operations might be as high as 0.5 percent of the initial cost per year. Pipe systems are deemed relatively free from repair and maintenance costs, but it is not unusual to incur R&M on pipelines.

## Labor

Labor can be a significant factor in operating costs. Some irrigation systems are inherently labor intensive. For example, hand move sprinkler pipe or siphon tube surface systems are very labor intensive. Center pivot systems, while not labor free, have low labor requirements. When calculating labor costs, the cost of benefits and workman's compensation must be included. If the labor is provided by the proprietor, the wage rate charged should be the opportunity cost of the proprietor's time.

## Trade-offs

Irrigation system design often involves trade-offs. For example, a more efficient engine or pump probably costs more. The question becomes one of how to compare the difference.

## Capital Recovery Factor

The capital recovery factor [CRF], also known as cost recovery factor, describes the repayment of a loan over a period of time in equal payments. As the loan principal is paid down, the interest per period decreases. If equal payments of principal were paid, with the interest for the period added, the periodic payment would change over the term of the loan. Most people prefer not to change the payment, so the CRF is used to calculate the periodic payment of principal and interest such that the payments are equal and the loan is just paid off with interest at the end of the number of periods of the loan. For example, if US$100 were borrowed at 1 percent per month for 10 months, the first payment would be US$11.00 (US$10 to principal and US$1 for interest). For the second month, the payment would be US$10.90 (US$10 to principal and US$0.90 for interest because only US$90 was owed for the period). However, applying the CRF, the monthly payment would be US$10.28 for each of the 10 months. The formula is shown below.

**Equations 12-3a and 12-3b**
*Capital recovery factor*

$$A = P \times CRF$$

$$CRF = \frac{i \times (i + 1)^n}{(i + 1)^n - 1}$$

where

| | | |
|---|---|---|
| A | = | amount of periodic payment |
| P | = | value of the amount now (at present time) |
| i | = | interest rate as a fraction per period |
| n | = | number of periods |

This formula looks a bit intimidating, but there is a variety of simple ways to get the answer. The simplest is to use a table of CRF values for times and interest rates. Most business calculators have the formula built in, and Excel® has a built-in function [PMT] that can be inserted into a cell. A table of CRF is shown in appendix A.

**Example 12-1**
*Comparative pump/engine efficiency costs*

Consider the case of an irrigation system operating 1,000 hours per year with a total head requirement of 200 feet and a flow rate of 750 gpm. Assuming an electric motor with an efficiency of 93 percent, how much could one afford to pay for a 1 percent increase in pump efficiency? Consider also that money costs 6 percent and electrical energy is $0.08/kilowatt-hour. To start, pump efficiency is 70 percent.

**Solution**

The water power required is $(200 \times 750) \div (3{,}960 \times 0.70) = 54.1$ hp $= 40.4$ kW.
Energy is $(40.4 \div 0.93) \times (1{,}000) = 43{,}441$ kWh.
Energy cost is $(43{,}441) \times (\$0.08) = \$3{,}475.28$.

Similarly, if the efficiency were 71 percent, the costs would be $3,423.60. The difference ($51.68) is the annual affordable investment amount. If the life of the pump is considered to be 20 years, CRF is 0.08718. By equation 12-3a, P = A ÷ CRF, so any cost difference less than $51.68 ÷ 0.08718 = $592.80 is positive. A difference from 70 to 75 percent is worth $2,698.33.

## Reduced Energy Alternatives

Once an energy source and energy supplier are selected, there is a defined rate for the energy used by the irrigation pumping system. Based on that rate, the annual costs of pumping become a function of the following variables:

- flow rate the pump is producing
- pressure/head the pump is producing
- overall pump and power unit efficiency
- hours of pumping per year

Anything that reduces any of the above variables, without changing the others, reduces the annual pumping cost.

### *Flow Rate*

If the pump produces less flow, the brake horsepower required and consequently the energy consumed decrease. However, the plants may still need the same amount of water. Therefore, with a few exceptions (noted below) the pump must operate more hours at the lower flow rate to apply the needed amount of water. The net result is usually no change in annual energy cost. One potential cost advantage occurs when lowering the brake horsepower reduces the electricity demand charge (per kilowatt) although the energy consumption charge (per kilowatt) remains the same.
Real and permanent savings can be realized if the reduction in flow occurs from practices that reduce the need for water. For instance, a more efficient irrigation system puts the same amount of water in the root zone with less volume of water pumped. The energy costs decrease with no adverse effect on the crop.

### Reduced Pressure

Reducing the pressure required by the irrigation system, without adversely affecting the application efficiency, reduces the energy consumption. A typical example is the change from high-pressure sprinklers to low-pressure sprinklers or spray nozzles. If there is no change in application efficiency, the energy consumption decreases. If the conversion also includes a more efficient irrigation system, the energy consumption improves even more: less energy per hour and fewer hours per year to apply the same volume of water to the root zone.

### Pump and Power Unit Efficiencies

Anything that makes the input power to waterpower conversion more efficient reduces energy costs. Put simply, it takes less energy to pump the same amount of water at the desired pressure. Choosing a more efficient pump or a more energy-efficient power unit are ways to achieve this end. However, the annual capital recovery costs of the more efficient power unit or pump should not exceed the savings in energy costs. The benefits of improving efficiency do not necessarily apply across different power units. For instance, electric motors have a much higher efficiency but may have a greater energy cost because electricity costs more per unit of energy.

The decision on whether to use diesel, natural gas, electricity, or other energy source is made first based on typical operating efficiencies and energy costs. Once that is done, the exact power unit can be selected using the efficiencies of different models as one point to consider. With IC engines, maintenance can significantly affect the operating efficiency. A poorly tuned IC engine can have a much lower efficiency and higher fuel consumption than a well-tuned engine doing the same job.

# Practice Questions

1. Compare the annualized cost of two pumps under the following set of conditions.

   Pump A — without VFD
   Pump B — with VFD

   **Given:**
   - Both pumps operate 1,000 hours per year for 10 years.
   - Base requirement of both pumps is as follows:
        25 percent of time 800 gpm, 260 feet of head
        50 percent of time 800 gpm, 210 feet of head
        25 percent of time 800 gpm, 170 feet of head
   - All three conditions have under-head pressure regulation. If there is excess pressure produced, the regulators burn off the excess pressure.
   - Estimated pump A efficiency is 0.79. It runs at 260 feet of head all the time.
   - Estimated pump efficiencies for pump B with VFD are the following:
        25 percent of time 800 gpm, 260 feet of head, 0.79
        50 percent of time 800 gpm, 210 feet of head, 0.81
        25 percent of time 800 gpm, 170 feet of head, 0.82
   - Motor efficiency is 0.92.
   - Money interest rate is 5 percent.

# Capital Recovery Factors

*Appendix*
# A

## Capital recovery factor [CRF]

$$CRF = [i(1+i)^n] \div [(1+i)^n - 1]$$

| Yrs, n | Rate, i (%) 1 | 2 | 3 | 4 | 5 | 6 | 7 | 8 | 9 | 10 | 11 | 12 | 13 | 14 | 15 |
|---|---|---|---|---|---|---|---|---|---|---|---|---|---|---|---|
| 1 | 1.01000 | 1.02000 | 1.03000 | 1.04000 | 1.05000 | 1.06000 | 1.07000 | 1.08000 | 1.09000 | 1.10000 | 1.11000 | 1.12000 | 1.13000 | 1.14000 | 1.15000 |
| 2 | 0.50751 | 0.51505 | 0.52261 | 0.53020 | 0.53780 | 0.54544 | 0.55309 | 0.56077 | 0.56847 | 0.57619 | 0.58393 | 0.59170 | 0.59948 | 0.60729 | 0.61512 |
| 3 | 0.34002 | 0.34675 | 0.35353 | 0.36035 | 0.36721 | 0.37411 | 0.38105 | 0.38803 | 0.39505 | 0.40211 | 0.40921 | 0.41635 | 0.42352 | 0.43073 | 0.43798 |
| 4 | 0.25628 | 0.26262 | 0.26903 | 0.27549 | 0.28201 | 0.28859 | 0.29523 | 0.30192 | 0.30867 | 0.31547 | 0.32233 | 0.32923 | 0.33619 | 0.34320 | 0.35027 |
| 5 | 0.20604 | 0.21216 | 0.21835 | 0.22463 | 0.23097 | 0.23740 | 0.24389 | 0.25046 | 0.25709 | 0.26380 | 0.27057 | 0.27741 | 0.28431 | 0.29128 | 0.29832 |
| 6 | 0.17255 | 0.17853 | 0.18460 | 0.19076 | 0.19702 | 0.20336 | 0.20980 | 0.21632 | 0.22292 | 0.22961 | 0.23638 | 0.24323 | 0.25015 | 0.25716 | 0.26424 |
| 7 | 0.14863 | 0.15451 | 0.16051 | 0.16661 | 0.17282 | 0.17914 | 0.18555 | 0.19207 | 0.19869 | 0.20541 | 0.21222 | 0.21912 | 0.22611 | 0.23319 | 0.24036 |
| 8 | 0.13069 | 0.13651 | 0.14246 | 0.14853 | 0.15472 | 0.16104 | 0.16747 | 0.17401 | 0.18067 | 0.18744 | 0.19432 | 0.20130 | 0.20839 | 0.21557 | 0.22285 |
| 9 | 0.11674 | 0.12252 | 0.12843 | 0.13449 | 0.14069 | 0.14702 | 0.15349 | 0.16008 | 0.16680 | 0.17364 | 0.18060 | 0.18768 | 0.19487 | 0.20217 | 0.20957 |
| 10 | 0.10558 | 0.11133 | 0.11723 | 0.12329 | 0.12950 | 0.13587 | 0.14238 | 0.14903 | 0.15582 | 0.16275 | 0.16980 | 0.17698 | 0.18429 | 0.19171 | 0.19925 |
| 11 | 0.09645 | 0.10218 | 0.10808 | 0.11415 | 0.12039 | 0.12679 | 0.13336 | 0.14008 | 0.14695 | 0.15396 | 0.16112 | 0.16842 | 0.17584 | 0.18339 | 0.19107 |
| 12 | 0.08885 | 0.09456 | 0.10046 | 0.10655 | 0.11283 | 0.11928 | 0.12590 | 0.13270 | 0.13965 | 0.14676 | 0.15403 | 0.16144 | 0.16899 | 0.17667 | 0.18448 |
| 13 | 0.08241 | 0.08812 | 0.09403 | 0.10014 | 0.10646 | 0.11296 | 0.11965 | 0.12652 | 0.13357 | 0.14078 | 0.14815 | 0.15568 | 0.16335 | 0.17116 | 0.17911 |
| 14 | 0.07690 | 0.08260 | 0.08853 | 0.09467 | 0.10102 | 0.10758 | 0.11434 | 0.12130 | 0.12843 | 0.13575 | 0.14323 | 0.15087 | 0.15867 | 0.16661 | 0.17469 |
| 15 | 0.07212 | 0.07783 | 0.08377 | 0.08994 | 0.09634 | 0.10296 | 0.10979 | 0.11683 | 0.12406 | 0.13147 | 0.13907 | 0.14682 | 0.15474 | 0.16281 | 0.17102 |
| 16 | 0.06794 | 0.07365 | 0.07961 | 0.08582 | 0.09227 | 0.09895 | 0.10586 | 0.11298 | 0.12030 | 0.12782 | 0.13552 | 0.14339 | 0.15143 | 0.15962 | 0.16795 |
| 17 | 0.06426 | 0.06997 | 0.07595 | 0.08220 | 0.08870 | 0.09544 | 0.10243 | 0.10963 | 0.11705 | 0.12466 | 0.13247 | 0.14046 | 0.14861 | 0.15692 | 0.16537 |
| 18 | 0.06098 | 0.06670 | 0.07271 | 0.07899 | 0.08555 | 0.09236 | 0.09941 | 0.10670 | 0.11421 | 0.12193 | 0.12984 | 0.13794 | 0.14620 | 0.15462 | 0.16319 |
| 19 | 0.05805 | 0.06378 | 0.06981 | 0.07614 | 0.08275 | 0.08962 | 0.09675 | 0.10413 | 0.11173 | 0.11955 | 0.12756 | 0.13576 | 0.14413 | 0.15266 | 0.16134 |
| 20 | 0.05542 | 0.06116 | 0.06722 | 0.07358 | 0.08024 | 0.08718 | 0.09439 | 0.10185 | 0.10955 | 0.11746 | 0.12558 | 0.13388 | 0.14235 | 0.15099 | 0.15976 |
| 25 | 0.04541 | 0.05122 | 0.05743 | 0.06401 | 0.07095 | 0.07823 | 0.08581 | 0.09368 | 0.10181 | 0.11017 | 0.11874 | 0.12750 | 0.13643 | 0.14550 | 0.15470 |
| 30 | 0.03875 | 0.04465 | 0.05102 | 0.05783 | 0.06505 | 0.07265 | 0.08059 | 0.08883 | 0.09734 | 0.10608 | 0.11502 | 0.12414 | 0.13341 | 0.14280 | 0.15230 |
| 40 | 0.03046 | 0.03656 | 0.04326 | 0.05052 | 0.05828 | 0.06646 | 0.07501 | 0.08386 | 0.09296 | 0.10226 | 0.11172 | 0.12130 | 0.13099 | 0.14075 | 0.15056 |
| 50 | 0.02551 | 0.03182 | 0.03887 | 0.04655 | 0.05478 | 0.06344 | 0.07246 | 0.08174 | 0.09123 | 0.10086 | 0.11060 | 0.12042 | 0.13029 | 0.14020 | 0.15014 |

*Appendix*
# B

# Answers to Practice Questions

## Chapter 1

1. A device to convert input energy into pressure and velocity to move water and pressurize systems.
2. (1) When the source is unpressurized and below where it will be used
   (2) If the pressurized source does not supply adequate pressure

## Chapter 2

1. (1) potential (pressure or head)
   (2) kinetic (velocity)
2. 0.433
3. $Q_1 \times Q_2 = A_1 \times V_1 = A_2 \times V_2$

## Chapter 3

1. (1) centrifugal — vertical turbine
   (2) centrifugal — submersible
   (3) centrifugal — jet
   (4) positive displacement
2. Vertical turbine
3. Submersible
4. Positive displacement

# Chapter 4

1. $Whp = \dfrac{H \times Q}{3{,}960}$

2. $Bhp = \dfrac{Whp}{E} = \dfrac{H \times Q}{3{,}960 \times E}$

3. $Bhp = \dfrac{550 \times 150}{3{,}960 \times 0.75} = 27.8 \text{ Bhp}$

# Chapter 5

1. $Q = \left[(0.24 \text{ in./day} \times 0.7) - 0.0\right] \times 12 \times \dfrac{43{,}560 \text{ ft}^2}{A} \times 0.623$

   $= 54{,}710 \text{ gal/day}$

   $= 54{,}710 \text{ gal/day}$

   If the system is available 24 hours per day, the flow rate needed is 38 gpm. However, if it is only available 12 hours in a 48-hour period, the flow rate needed is $\dfrac{38 \times 48}{12} = 152$ gpm

2. $H_f = 0.86 \text{ psi}/100 \text{ ft} \times 300 \text{ ft} = 2.58 \text{ psi} = 5.96 \text{ ft}$

   $H_p = 40 \text{ psi} = 92.4 \text{ ft}$

   $H_L = 15 \text{ ft}$

   $H_v = \dfrac{(4.98)^2}{2 \times 32.2} = 0.39 \text{ ft}$

   $TDH = H_f + H_p + H_L + H_v = 5.96 + 92.4 + 15 + 0.39 = 113.75 \approx 114 \text{ ft}$

3. When the pressure in the pump falls below the vapor pressure of the water (at that temperature), liquid flashes to vapor forming vapor bubbles. Later in the pump, when the pressure rises above the vapor pressure, the bubbles collapse.

# Chapter 6

1. First (1)
2. Second (2)
3. Third (3)
4. In series, head is additive and flow is the same.
5. In parallel, flow is additive and head is the same.

# Chapter 7

1. (1) total dynamic head [TDH]
   (2) design flow rate
2. Any five of the following: suction pipe friction loss, suction lift, suction entrance loss, discharge pipe friction loss, discharge lift, sprinkler operating pressure, miscellaneous fittings loss
3. As altitude increases, there is less atmospheric pressure thus reducing NPHSa.
4. As temperature of the water goes up, so does the vapor pressure thus reducing NPHSa.
5. Friction loss on the suction side reduces NPHSa, and NPSHa must be greater than NPHSr. Hence, larger suction sizes reduce friction loss.
6. Any of the following: flow, head, efficiency, horsepower, NPHSr, speed, impeller diameter, maximum solids size, suction size, discharge size, number of vanes
7. H = 219 ft
   Efficiency = 81%
   Bhp = 68 hp
   NPSHr = 10+ feet

# Chapter 8

1. Suction lift and elevation lift
2. Suction friction loss, suction entrance losses, pressure head, and friction head
3. The pump operating point is where the system curve and pump curve intersect. Normally that is only one point.

# Chapter 9

1. Any of the following: low maintenance, simple operation, long life, easy to automate
2. (1) easier to move
   (2) speed is variable
3. NPPPC allows comparison of pumping costs of various types of power plants.

# Chapter 10

1. (1) pressure regulation with valves
   (2) power unit speed control
   (3) on/off control
2. VFDs change the frequency of the power to an electric motor thereby changing the speed.
3. (1) insulation stress
   (2) harmonic distortion
   (3) bearing currents
   (4) resonant frequencies
4. Flat

# Chapter 11

1. If conditions require reducing pressure, a VFD can reduce pressure by lowering the pump speed and thereby reduce pump power. The alternative of pressure regulation with a valve causes friction loss and does not reduce power as much as it would with a VFD.
2. Bearing failure or casting breakage
3. Excess heat causes insulation to break down and can destroy the motor.

# Chapter 12

The VFD allows pump B to use less power when slowed down to meet head conditions, but at all times the VFD is 97 percent efficient.

Pump A
  a. Bhp = (800 × 260) ÷ (3960 × 0.79) = 66.5 Bhp
  b. $P_{IN}$ = Bhp/$E_{PP}$ = 66.5 ÷ 0.92 = 72.3 hp
  c. Converting to kW, 72.3 hp × 0.746 kW/hp = 53.9 kW
  d. Running 1,000 hours per year, it uses 53.9 kW × 1,000 h = 53,900 kWh.
  e. At $0.10 per kWh, cost is $5,390.

Pump B (assuming the VFD can match the operating condition)
  a. 25 percent of time, kW is 0.746 × (800 × 260) ÷ (3,960 × 0.92 × 0.79) = 53.9 kW
  b. 50 percent of time, kW is 0.746 × (800 × 210) ÷ (3,960 × 0.92 × 0.81) = 42.5 kW
  c. 25 percent of time, kW is 0.746 × (800 × 170) ÷ (3,960 × 0.92 × 0.82) = 34.0 kW
  d. Accounting for VFD efficiency, power consumption for the year is
     [(53.9 × 250) + (42.5 × 500) + (34.0 × 250)] ÷ 0.97 = 44,562 kWh.
  e. At $0.10 per kWh, cost is $4,456.

On the surface, using the VFD would save $934 per year. The CRF for 5 percent and 10 years is 0.12950, so the present value of the energy savings is $934 ÷ 0.12950 = $7,212. If the VFD pump system for pump B can be bought for less than $7,212, it is the better alternative. Remember that it likely will require both a VFD and an upgraded motor.